VEDIC SCIENCE AND TECHNOLOGY

DR. M R GOYAL

गरुड

Published by
Garuda Prakashan Private Limited
Gurugram, Bharat

www.garudabooks.com

First published in India 2022

ISBN: 978-1-942426-44-8

Cover Design: Jitender Nigam

Printed in India

Contents

Acknowledgements

The author is grateful to his son, Puneet Goel, daughter Geetanjali and granddaughter, Nandini, for their assistance in preparing and editing the manuscript.

He is also grateful to Mr. Sankrant of Garuda Prakashan, without whose encouragement this book would not have come to light.

Preface

Several books are available in the market on the topic, "Science and Technology in Ancient India". This book is unique in the sense that it brings out those aspects which depict more advanced Science and Technology in ancient India than in modern times. The book has been divided into four parts.

The first part, containing two chapters, gives a strict scientific proof of the fact that science in ancient India was really developed, at least to the level of modern science, and that it is not a matter of conjecture only.

The second part of the book, containing nine chapters, talks about some of the technologies found in the ancient Sanskrit texts which were really much more advanced than the modern technology. Imminent research can be taken up based on the information/ knowledge contained in these chapters. It may lead to a quantum leap in the advancement of science and technology.

The third part, containing eight chapters, depicts some advancements which are not imminently implementable, but give a glimpse of futuristic technologies, something that was already achieved in ancient India. If the ancient Science and Technology books that are presently not available in India are located and deciphered, it may result in further rapid advancement in science and technology.

The fourth part, containing two chapters gives dating of ancient Indian history and a bit of a glimpse of pre-historic history.

PART-I

Strict Scientific Support

CHAPTER 1

Standard Units of Length and Time

All scientifically advanced civilisations have standard units of measurement. The standard units are maintained in the form of scales etc., which are proven not to vary with change in temperature etc.

In the excavation season of 1930-31 at Mohenjo-Daro (now in Pakistan) in the Indus Valley, the archeologists found one such scale carved on a piece of shell (which does not expand or contract with variation in temperature) [Michel Danino, *The Lost River*]. It had precisely-spaced eight divisions, each division being equal to 6.7056 mm (0.264 inches), with a circle and a dot, marked at five division internals. The five divisions of the scale, work out to 33.528 mm (1.32 inches). This measure is widely known as an Indus Inch. The dating of the scale indicated 3500 BCE or earlier.

Another ancient scale, known as Gudea's Rule (2175 BCE) was found in Sumeria and is preserved in the Louvre, Paris. Surprisingly, each division of that scale is 0.66 inches, which is exactly equal to half an Indus Inch. This unit of 0.66 inches, is known as Sumerian 'shusi', which means finger or *angula.* It shows that the standard unit of length in India as well as in Sumeria was an angula, equal to 0.66 inch (16.764 mm). It also shows that the Indo-Sarasvati Civilisation had spread to Sumeria by 2200 BCE.

Relationship of units of length measurement, is documented in Kautilya's *Arthshastra,* a well-known treatise on politics, authored by Chanakya, the political mentor of Chandra Gupta Maurya. In the *Arthshastra,* two types of Dhanush are mentioned, as units of length. An ordinary Dhanush consisted of 96 angula, and a Garhaspatya Dhanush, used for measurement of roads and distances, consisted of 108 angula.

Further, a Yojana consisted of 8,000 (garhaspatya) Dhanusha.

Exact measurement of the Garhaspatya Dhanush would thus be $16.764 \times 108 = 1810.512$ mm

$= 1.810512$ metres.

and the length of an ordinary Dhanusha,

would be $16.764 \times 96 = 1609.344$ mm

$= 1.609344$ metres.

Two engravings on a wall of the temple of Tirupattkali (12th century CE) near Kanchipuram in South India depict two scales - one measuring 7.24 metres in length, with markings dividing the scale into four equal parts, each measuring 1.81 metres, which is equal to a Garhaspatya Dhanush; and the second scale measures 5.69 metres in length, and has markings dividing the scale into four equal parts. It may be observed that the length of the second scale is equal to the circumference of a circle with its diameter equal to a Garhaspatya Dhanush i.e. 1.81 metre.

Relationship with Modern Mile

It is a well-known fact of history that 'Mile' was earlier known as 'Mille', which literally means a 'thousand'. Thus, a 'Mile' or 'Mille' consisted of a 'thousand' units. But it could not be found so far as to what that unit was.

By dividing the length of a 'Mile' by 1000, we can find that unit. It works out to be exactly equal to 1.609344 metres; that is an ordinary Dhanush. This relationship establishes, beyond any doubt, that the standard length of an angula in ancient civilisations of India, as well as Sumeria, was 16.764 mm.

It is surprising to note as to why these relationships between various precise scales have not been established so far, although the precise scales were found about 90 years ago. The main reason appears to be some prejudices.

One such prejudice is that the 'angula' measured to be equal to 17.63 mm. The basis of which is totally unscientific. This length was calculated by measuring the thickness of fingers of several men and finding the average thickness of finger. This should have been discarded long ago, when the standard scales were found in the archaeological excavations of Indo-Sarasvati civilization and the Sumerian, civilization.

The standard length of 'angula' must be taken as 16.764 mm, instead of 17.63 mm.

Standard Unit of Time

In Vishnu Purana (Book 1, Chapter 3, Shlokas 8, 9), it is narrated that:

15 Nimishas = 1 Kashtha

30 Kasthas = 1 Kala

30 Kalas = 1 Muhurta

30 Muhutras = 1 Ahoratra (Day and Night).

The basic unit of time, taken for scientific calculations was a Nimisha.

In the famous book on Astronomy *Surya Siddhanta* [Ref: Chapter 1, Shloka 12], Ahoratra is qualified as Nakshatra Ahoratra i.e. Sidereal day, indicating that for Astronomical and other scientific calculations, the Ahoratra was taken to be equal to Nakshatra Ahoratra, i.e. the time taken by the stellar constellations, to complete one revolution of the Earth. As the stellar constellations complete 366.25 revolutions in a year (365.25 days), a Nakshatra Ahoratra is slightly shorter than the normal day of 24 hours.

Thus, a Nakshatra Ahoratra = $24 \times \frac{365.25}{366.25}$ hours

= 23 hours 56 minutes and 4.1 seconds

= 86,164.1 seconds

Thus, one Nimisha = $\frac{86,164.1}{30^3 \times 15}$ second

= 0.2127508 second

Thus, the standard unit of time in ancient India was a Nimisha = 0.2127508 second.

CHAPTER 2

Speed of Light

In the commentary on Rig Veda (Mandal 1, Sukta 50, Mantra 4), Sayanacharya (14th Century CE) had cited an ancient Sanskrit Shloka, which reads:

योजनानां सहस्रे द्वे द्वे षते द्वे च योजने।
एकेन निमिशार्धेन क्रममाण नमोडस्तु ते।।

Meaning: "Salutation to thee (the sun) who approacheth 2202 Yojanas in a Nimishardha".

The word 'approacheth' indicates that the distance per unit time, is the speed of light, because the sun approacheth through its rays, at the speed of light.

As pointed out in the last chapter.

One Yojana = 8,000 (Garhaspatya) Dhanushas and one Garhaspatya Dhanusha = 1.810512 metres

Therefore, One Yojana = 8,000 ×10810512 metres

= 14.484096 km

Also, one Nimishardha = ½ × 0.2127508 = 0.1063754 sec.

Thus, the speed of light works out to be equal to

$$\frac{2,202 \times 14.484096}{0.1063754} \text{ km/sec}$$

= 2.998×10^5 km/sec which is precisely equal to the speed of light, as calculated in modern science.

It could not have been possible to measure the speed of light so accurately without the use of very sophisticated scientific instruments.

It clearly establishes that science and technology in ancient India, were highly advanced.

PART-II

Innovative Ideas for Imminent Implementation

CHAPTER 3

Microwave Weapon

In an ancient Sanskrit book *Brihadviman Shastra,* the range of the spectrum of electromagnetic waves, emanating from the Sun, has been given.

The highest frequency of the waves has been mentioned as महाक्षोणीत्रयम्. The dictionary meaning of महाक्षोणी is "One followed by 24 ciphers" i.e. 10^{24}. Thus, the highest frequency observed by the ancient Indian scientists, was 3×10^{24} cycles per unit of time. The standard unit of time, used for scientific calculations in ancient India was 'निमिष', which has been shown to be equal to 0.2127508 seconds.

Thus, the highest frequency of the electromagnetic waves, observed by the ancient Indian scientists, works out to $\frac{3\times10^{24}}{0.2127508}$ cycles/sec = 1.4×10^{25} Hertz.

As per the latest observations in modern science, the highest frequency of electromagnetic waves, observed so far is of the order of 5×10^{25} Hertz. This frequency was observed by the latest and most powerful 'Fermi telescope', launched in a satellite, in 2008. Prior to that, the highest frequency observed in modern science was much lower than 1.4×10^{25} Hertz, observed by the ancient Indian scientists.

It shows that the ancient Indian scientists, had a telescope, at least as powerful as the 'Fermi telescope'.

The lowest frequency of electromagnetic waves, as observed by the ancient Indian scientists, was 90 Hertz, which also tallies well with the lowest frequency in modern science.

It has been further mentioned in the ancient Sanskrit book that the whole spectrum, from the highest frequency to the lowest was divided by the ancient Indian scientists, into three main divisions.

The three main divisions were further divided into very small sub-divisions in ancient India. In fact the scientists of the "Valmiki school of fundamental physics", had divided the first 'high frequency' division into 50 million sub-divisions, the middle division, into 8,000 sub-divisions and the lower frequency division into 107 sub-divisions. All these sub-divisions were further divided into 10 parts each.

Further, it has been mentioned that if three specific frequency waves, in the second division are combined, then so intense heat, is produced that the (enemy) plane can be burnt. The three specific frequencies have been indicated in the book. This weapon can be a very effective substitute of a laser weapon, which is still being developed.

Experiments need to be conducted on the basis of the three specific frequencies indicated in the ancient book.

The relevant verses are given below:

तदुक्तमृतु कल्पे -
महाक्षोणित्रयं पश्चात् कोटीनामेकविंशतिः।
लक्षाणां पंचाशत् पश्चात् सहस्राणां तु षोडश ॥
पश्चादेकोनविंशत् संख्याकाः सूर्यमरीचयः।
प्रसरंति विशेषेणादितेर्ग्रीष्माख्य गर्भतः ॥
तेषां वर्ग विभागस्तु वाल्मीकि गणिते क्रमात्।
पंचकोट्यष्टसहस्रं सप्तोत्तरशतं स्मृतम् ॥

तेषामेकैकवर्गेथ विभागा दशधा कृताः।
तेषु द्वितीयवर्गस्थ विभागेषु यथाक्रमम् ॥
त्रिपंचदशमौष्म्यांशु मेलनं ग्रीष्ममध्यमे ।
यदा भवति ग्रीष्मोष्मा कूर्मान्तं व्याप्यते स्वयम् ॥
पश्चात् कच्छप प्रम्लोच शक्त्याकर्षणतः क्रमात् ।
कुलकाख्या जायते काचिच्छक्तिर्ज्वलन वत्स्वतः ॥
तत्संयोगो यदि भवेद् व्योम्नि यान पथि क्रमात् ।
भस्मीकृतं भवेद् व्योमयानमत्यन्त शीघ्रतः ॥
तदपायविनाशार्थं कुटिणीशक्तियन्त्रकम् ।
संस्थापयेद् यानकण्ठप्रदेशे सम्प्रदायतः ॥

Meaning: (It is stated in the book *Ritu-Kalpa*) Three "Mahakshonis" i.e. 3 x10^24, then 21 crore 50 lac, then 16 thousand, then 19: Sun-rays (electromagnetic rays) with these numbers i.e. frequencies, emanate specially from the Sun's hot core. In the Valmiki School of Maths (Fundamental Physics), these have been divided respectively into 5 crore, 8,000 and 107 divisions (i.e. the electromagnetic waves lying between frequencies from 3 X 10^24 to 21.5 X 10^7 have been divided into 5 crore divisions, waves lying between 21.5 X 10^7 to 16,000, have been divided into 8,000 divisions, and the waves lying between 16,000 to 19 have been divided into 107 divisions).

In each of these categories, the divisions have been further subdivided into 10 sub-divisions each. Out of these, when the waves of 3rd, 5th and 10th sub-divisions of the (8,000) divisions of the second category of waves, intermingle, then automatically, heat with extreme temperature is generated. Then due to the beats (कुलक्क) generated by intermingling of these frequency waves, a blaze (violent flame) like power named Kulak (कुलक, कुलक्क) is generated. If such a phenomenon happens in the path of an aeroplane, then the aeroplane can catch fire and get burnt very quickly.

To safeguard against this calamity, the instrument named "Kuntini" (कुंटिणी शक्ति-यन्त्र) is installed in the neck-region of the aeroplane.

Further clarification on the wave frequencies, is given in the following verses:

नारायणोपि -

ग्रीष्मोष्म किरण वर्ग विभागेषु यथाक्रमम्।
द्वितीयवर्ग किरणाः सन्त्यशीति सहस्रशः II
तेष्वष्टत्रिदश संख्याकांशवोत्यन्तमूष्मकाः।

Meaning: "Narayana Scientist also (says),

Out of the different divisions of the electromagnetic waves, the waves of the second category, are divided into eighty thousand subdivisions. Out of these, the waves of 8th, 3rd, and 10th subdivision are extreme heat-producing."

लल्लोपि -

ग्रीष्मोष्म किरण वर्ग विभागेषु यथाक्रमम्।
द्वितीयवर्ग द्वात्रिंशद् विभागांशुषु क्रमात् II
पंच त्रि दश संख्याकाः किरणा ऊष्म्यरूपिणः।

Meaning: "Lall Scientist also (says),

Out of the different categories of the electromagnetic waves, the waves of the 5th, 3rd and 10th sub-divisions of the (first) thirty-two divisions of the 2nd category, are much heat-producing."

It may be observed that the beats produced by the intermingling of waves of 3rd, 5th and10th sub-divisions, would be exactly similar to the beats produced by the intermingling of waves of 3rd, 8th and 10th sub-divisions. Therefore, there is no contradiction between the statements of Scientist Narayana and Scientist Lall. The only difference is that while Narayana talks of the subdivisions of only the first division of the second category waves, Lall talks of the subdivisions of the first thirty two divisions of the second category waves.

The above narration indicates the existence of not only a Micro-wave superweapon, but also a counter-weapon (कुंटिणी शक्ति यन्त्र) to nullify the destructive effect of the Micro-wave weapon. In the book 'यन्त्रसर्वस्व', a brief description of कुंटिणी शक्ति यन्त्र is given in six pages. However, the technical terminology used therein is difficult to decipher. Probably, a strong electromagnetic field is created which absorbs (पिबति) the beats (कुलक्क) produced by the intermingling of the three microwave frequencies. But the process to create the electromagnetic field is quite complicated and difficult to understand (decipher), and is a subject of research.

CHAPTER 4

Solar Cell

In the ancient Sanskrit book, *Brihadviman Shastra,* an advanced solar power device has been described. The device is named as "सूर्यशक्त्यपकर्षण यन्त्र", meaning an instrument to convert solar power. In this instrument, instead of the mica layer used in the modern solar cell, a liquid paste, prepared by blending and heating several components in a crucible, is used, thus, making it a hybrid solar cell combining the advantages of the modern silicon-based cells and the organic solar cells. The relevant Shlokas from the book are given below:

स्फाटिक मंजुलफेनसुवर्चान् सैकतपारदगरदकिशोरान्।
गंधक कर्बुर प्राणक्षारान् रविशशिपंचमुखामरपंकान्।।
रविवसुदिङ्नक्षत्रविभागान् वेदानलसागरवसु भागान्।
सायकपादपभूतविभागान् वसुमुनिनिधिनेत्रविभागांशान्।।

In the first two lines of the shlokas, the names of the substances to be used are given and in the next two lines, their proportions (percentages by weight) are given. Two of the substances are :

1) Powder of quartz - 12% and 2) Powder of sand - 4%

Clearly, these are silicon-based substances, containing more silicon than mica. Some of the remaining substances are inorganic and others are organic substances. The hybrid solar cells, thus created, may have several advantages over the modern solar cells:

Three other components:

Sodium Carbonate 27%

Samudra Jhag (समुद्र झाग) 10% and

Ammonium Chloride 5%

are added to substantially lower the melting point of quartz and sand. Thus, on heating these 5 substances, a very cheap substitute of a silicon-based solar cell can be obtained.

The other components can then be added one-by-one, to obtain a hybrid solar cell, with higher efficiency.

Mercury (3%) is perhaps added to create a very thin conducting layer, as a substitute for the thin wire-mesh, in the modern solar cells. Laboratory experiments, on these lines need to be carried out.

The other components are listed below :

Tropical Fig (Fruit) 8%

Nux Vomica 7%

Tail Parni (तेल–पर्णी) 8%

Sulphur 5%

Shampoo Ginger/धतूरा 1% and

Rudraksha (रुद्राक्ष) 10%

The amount of liquid paste to be painted over a surface of 20 angula x 80 angula is given as one cubic angula, which gives the thickness of layer as 10.5 microns. The thickness of silica layer in modern day solar cell is 40 microns. As the efficiency is more with smaller thickness of this layer, the efficiency of the ancient solar cell would be much higher than that of the modern solar cell.

CHAPTER 5

Metallurgy and Mortar

In India, the technology related to metallurgy had been developed earlier than 4000 BCE. Several idols of copper, dated 4000 BCE or earlier, have been found in archeological excavations in India. Idols and artefacts of other metals and alloys, like brass and bronze, have also been found. High quality stainless steel was also manufactured in ancient India. In the ancient Sanskrit book, *Brihadvimana Shastra,* the manufacture of 16 different types of light weight, heat resistant and fire proof alloys has been mentioned.

For the extraction of metals from ores and the manufacture of alloys, some substances,which substantially lower the melting point of the metal are mentioned. One such substance is टंकण meaning सुहागा.. or borax.

With substantially lower melting points, the metals and alloys could be manufactured in small-scale cottage industries, thus avoiding large scale pollution caused by big industries. Otherwise as well, environmental pollution and global warming would be substantially lower due to the low melting points of the metals. Consumption of energy would also be much lower, and the cost of manufacturing would also be substantially less. With a shift from the big industries to the small-scale industries, there would be several indirect benefits too, with back-to-village call, for instance. Research in this direction is very important and essential, in view of the present scenarios of global warming and threat of pandemics looming large over the world.

Immediate action in this respect needs to be taken. A vehement opposition to such ideas is expected from the big industries, which will have to be overcome for the welfare of the nation and humanity as a whole.

Mortar:

In modern day construction, the mortar used consists of two components viz. sand and cement. Sometimes, this is substituted by lime. The usage of both these components is an environmental hazard. In ancient India, mortar consisted of a small amount of lime mixed with several other organic components. The composition is given in the ancient book, *Maya-Mata* (मय मत), and has been reproduced in a chapter in the book, "Pride of India". The readers may refer to any of the two books named above for details. The mortar so obtained was much stronger than the cement-sand mortar. Moreover, the thickness required was less than a millimetre, as against ten millimetres of cement-sand mortar.

Another much simpler method of preparing mortar, as practised in ancient India, is given below :-

"Mix equal parts of fresh sand and lime, and leave in a pond of water for three days. Taking it out, sprinkle a liquid slurry of गुड़ over it (molasses mixed with water can also be used). Then, consolidate the mixture by thrashing. A thin layer of slurry thus produced is to be pasted on the surface of the bricks. Even this simple mortar was stronger than the present day cement-sand mortar. [Ref: प्राचीन भारत में विज्ञान एवं प्रौद्योगिकी, p 45].

In the context of present-day environmental pollution and global warming, there is urgent need to switch over to the mortar used in ancient India. If necessary, there can be some modifications in order to reduce the number of organic components, after testing in laboratories. Of course, there will be a lot of resistance from cement industries to this switch-over as well. The ancient mortar can easily be prepared in village-level, small-scale cottage industries.

CHAPTER 6

Cure of Heart Disease and Jaundice

There is a Mantra is Rig-Veda, which reads:

"उद्यन्नद्य मित्रमह आरोहन्नुत्तरां दिवम्।
हृद्रोगं तु मम सूर्य हरिमाणं च नाशय।।"
(ऋ. मंडल 1 सूक्र 50 मंत्र 11)

Meaning:

"O great friend Sun, now arising from the horizon and ascending towards the upper sky, remove my heart disease and the jaundice".

The mantra clearly indicates that heart-disease and jaundice can be cured by exposure to the rays of the rising Sun, for some days.

It may be noted in this context that the modern medical science has recognized that jaundice can be cured by exposure to the rays of the morning Sun. However, regarding the cure of heart-disease, the modern medical science has not come out with a similar cure.

When it is recognized that the Mantra is true in respect of one disease, it is very likely that it would also be true, for the other disease mentioned therein. It is, however, possible that the cure given in the mantra may be fully effective in some types of heart disease but may not be so effective in other types.

Experiments must be conducted by medical experts - allopathic, ayurvedic naturopathic and homeopathic - to check the efficacy of the cure mentioned in the Veda-mantra. Patients suffering from different types of heart-diseases may be divided in two equivalent groups. One group may be exposed to the rays of the rising Sun, daily. All other treatments given to the two groups should be the same. The difference in improvement in the two groups be observed after every few days.

These would be harmless experiments as the group to be exposed to the Sun-rays will not be deprived of the other treatments. However, the results may prove to be of immense benefit for the heart patients all over the world.

CHAPTER 7

Gene-Therapy for Infantile Seizures

In *Bhagavata Purana* [Skandha 10, Chapter 6], wherein the description of Putana is narrated, it is clearly mentioned that Putana was an infantile seizure (बाल–ग्रह). In *Sushruta Samhita,* a well-known treatise on Ayurveda, also, Putana is named among nine types of infantile seizures, mentioned therein. In fact, there are four types of Putana, out of the nine types of seizures, described therein, viz. Putana, Andha Putana, Shita Putana, and Mukhamandika Putana. The following shlokas of उत्तर–तन्त्रम्, Chapter 27, are relevant:

स्कन्दग्रहस्तु प्रथमः स्कन्दापस्मार एव च।
शकुनी रेवती चैव पूतना चान्ध–पूतना।।
पूतना शीतनामा च तथैव मुखमंडिका।
नवमो नैगमेषश्च यः पितृग्रहसंज्ञितः।।

Meaning: "The first of the nine infantile seizures, is Skanda, then Skandapasmara (epilepsy), Shakuni, Revati, Putana, Andha Putana, Shita Putana, Mukha mandika Putana and Naigamesha which is also called Pitrigraha (a genetic disease inherited from father). The symptoms of these seizures are given in chapter 35 of उत्तर–तन्त्रम्

It may be noted that the description of Putana given in Bhagavata Purana, is a detailed narration of the symptoms of Mukhemandika Putana, given in Sushruta Samhita, as under:

अलंकृता रूपवती सुभगा कामरूपिणी।
गोष्ठमध्यालयरता पातु त्वां मुखमंडिका।।

Meaning: "may the ornamented, beautiful, fortunate enchantress, abiding in the cowsheds, Mukhamandika Putana, protect you".

Compare it with the detailed description given in Bhagavata Purana [Skandha 10, Chapter 6]:

तां केशबन्धव्यतिषक्तमल्लिकां बृहन्नितम्बस्तन कृच्छ्रमध्यमाम्।
सुवाससं कम्पितकर्ण भूषणत्विषोल्लसत् कुंतलमंडिताननाम्।।
वल्गुस्मितापांग विसर्गवीक्षितैर्मनो हरन्तीं वनितां व्रजौकसाम्।
अमंसताम्भोजकरेण रूपिणीं गोप्यः स्त्रियं द्रष्टुमिवागतां पतिम्।।

Meaning: "She (Putana) had jasmine flowers interwined in her hair-lock, had robust hips and breasts and a slim waist, was wearing beautiful clothes, her face was decorated by tressles of hair glittering with the light radiated by her shaking ear-rings. She was bewitching the village-folk by her lovely, smiling looks cast from her eyes. The cowherdesses thought that she was a beautiful woman with lotus hands, who had come to look for her husband".

In Bhagavata Purana also, several other names of infantile seizures (अर्भक–ग्रहाः) and other child-diseases have been mentioned and also gene-toning therapy against the genetic diseases, including the infantile seizures. The relevant Shlokas are given below:

बालं तं तस्या उरसि क्रीड़न्तमकुतोभयम्।
गोप्यस्तूर्णं समभ्येत्य जगृहुर्जातसंभ्रमाः।।
यशोदारोहिणीभ्यां ताः समं बालस्य सर्वतः।
रक्षां विदधिरे सम्यग्गोपुच्छ भ्रमणादिभिः।।
गोमूत्रेण स्नापयित्वा पुनर्गोरजसाभ्रकम्।
रक्षां चक्रुश्च शकृता द्वादशांगेषु नामभिः।।
गोप्यः संस्पृष्टसलिला अंगेषु करयोः पृथक्।
न्यस्यात्मन्यथ बालस्य बीजन्यासमकुर्वत।।
डाकिन्यो यातुधान्यश्च कूष्मांडा येऽर्भकग्रहाः।

भूतप्रेतपिशाचाश्च यक्षरक्षोविनायकाः ।।
कोटरा रेवती ज्येष्ठा पूतना मातृकादयः ।
उन्मादा ये ह्यपस्मारा देहप्राणेन्द्रियद्रुहः ।।
स्वप्नदृष्टा महोत्पाता वृद्धबालग्रहाश्च ये ।
सर्वे नश्यन्तु ते विष्णोर्नाम ग्रहण भीरवः ।।

Meaning: "The perplexed cowherdesses, quickly going near her (the de ceased Putana), caught hold of the baby Krishna playing fearlessly on her breast. Along with Yashoda and Rohini, they performed ceremonies for the all-round protection of the child, by revolving cow-tail etc.

Bathing the child with cow-urine and then with cow-dust and smearing cow-dung on the twelve body parts, uttering their names they protected the child. The cowherdesses after sprinkling water on the body parts and the hands separately and throwing it on their body did gene-toning therapy (बीजन्यास) of the child.

The imps, demonesses, disease-creating demons, the seizures of children, ghosts, evil spirits, fiends, yakshas, Rakshasas, Vinayakas, Kotara, Revati, Jyeshtha, Putana, Matrikas (genetic diseases inherited from mother), manias, epilepsies which harm the body, life-force and sense-organs, the nightmares, deliriums and seizures of the elderly and children - may all these be destroyed as these are frightened at the utterance of Vishnu's name".

It is evident from the above narration that babies can be protected from the infantile seizures and other horrible diseases of childhood, by occasionally bathing them with cow-urine and cow-dust and then smearing the body-parts with cow-dung and giving a shower thereafter.

By the use of cow-urine and cow-dung in the manner given herein the paedriatic diseases and infant mortality rate can be reduced to a large extent.

CHAPTER 8

Concept of Homeopathy

Before Hahnemann propagated it in the modern times, the principle of Homeopathy finds enunciation in the *Bhagavata Purana*, about 5,000 years ago. The relevant shloka in the *Bhagavata Puran* reads.

आमयो यश्च भूतानां जायते येन सुव्रत।
तदेव ह्यामयं द्रव्यं न पुनाति चिकित्सितम्।।

[स्कन्ध 2, अध्याय 5]

Meaning: "Whichever disease is created in the living beings, by a substance, does that very substance, therapeutically prepared not cure that very disease?" This is the basic principle of Homeopathy.

Interestingly this principle is applied to deduce the Philosophy of "निष्काम कर्म–योग", in the next shloka, which reads :

एवं नृणां क्रिया योगाः सर्वे संसृतिहेतवः।
त एवात्मविनाशाय कल्पन्ते कल्पिताः परे।।

Meaning: "Similarly, all activities of human beings are a cause of bondage with the world but those very activities, dedicated to God, lead to liberation".

The therapeutics of preparing medicines by repeated dilutions is indicated in *Charaka Sanhita* (चरक–संहिता), the well-known treatise on Ayurveda. A shloka for the cure of a type of sinusitis reads:

सिद्धं ससैंधवं तैलं दशकृत्वोऽणु तत्स्मृतम्।
स्निग्धमास्थापनैर्दोषं निर्हरेद्वातपीनसे।।

Meaning: "It is known that oil prepared by adding सैंधा (common salt) and diluted iteratively ten times, administered to the patient suffering from sinusitis due to वात–दोष, after greasing him with energy-boosting substances, cures that disease".

This therapeutic preparation is similar to preparing 10th potency of Natrum mur, in present day Homeopathy, except that the dilutions were made in oil, instead of alcohol.

To some extent, the process of iterative dilutions, is still prevalent in our house hold cures. For example, for the cure of chronic constipation, त्रिफला (a mixture of हरड, बहेड़ा and आँवला) is immersed in a pot of water and kept overnight. In the morning, the diluted solution from the upper portion of the liquid is administered to the patient. To the remaining portion, more water is added to prepare the second dilution (potency) and the upper portion administered to the patient, next morning.

This process is repeated for a few days, so that higher and higher potencies are administered to the patient on successive days. This is a very effective and tried cure of chronic constipation.

CHAPTER 9

Embryo Transplantation

It is mentioned in *Harivansh Purãna* [Book 2 chapter 4] and *Bhãgavata Purãn* [Skandh 10, Chapter 2] that when Balarãm was conceived by Devaki, his embryo was transplanted to the womb of Rohini. The relevant shlokas are:

"आपन्नं सप्तमं गर्भं तां निनायाथ रोहिणीम्।
अर्धरात्रे स्थितं गर्भं पातयन्ती रजस्वला।।
निद्रया सहसाविष्टा पपात धरणीतले।
सा स्वप्नमिव तं दृष्ट्वा गर्भं निःसृतमात्मनः।
अपश्यन्ती च तं गर्भं मुहूर्तं व्यथिताऽभवत्।। [ह.पु.]
गर्भे प्रणीते देवक्या रोहिणीं योगनिद्रया
अहो विस्रंसितो गर्भ इति पौरा विचुक्रुशुः।। [भा.पु.]

Meaning: "When the seventh child was conceived, its embryo was transplanted into Rohini's womb. When Devaki was discharging the embryo at mid-night she menstruated and suddenly overpowered by (induced) sleep, fell down on the ground. After seeing her discharged embryo as if in a dream, and then not seeing it, she felt pained for a while. On the embryo having been transplanted, from Devaki to Ronini's womb, by the process of induced sleep (Yoga Nidrã), the people cried, "Alas! The pregnancy has been miscarried".

From the above narration, it is clear that in Mahabharata era, operation for embryo - transplantation was known and that the surgeons used the process of induced sleep instead of chemical anaesthesia, during major surgical operations.

CHAPTER 10

Removal of Body-Hump by Physiotherapy

In *Bhāgavata Purāna* and *Harivansha Purāna*, an incident of a hunch-backed girl (Kubjā) being cured of her hump by Krishna using physiotherapy technique is described. The process of the technique adopted can be understood by combining the narrations given in the two holy books, and dovetailing the two narrations.

When the king Kansa had invited Nanda and other Gopas along with Krishna and Balarama from Vraja to Mathura with the objective of getting both of them killed in a wrestling bout, the two young boys, Krishna and Balarama, were taking a stroll on the royal road in Mathura. While walking along the royal road, they saw a hump-backed girl (Kubjā) with an anointment pot. She was probably going to the royal palace to provide the anointment to Kansa and his family members. Impressed by the loveliness, delicacy, sweetness, smile, talks and glances of the two young boys, she stood laughing and offered anointment suitable for their bodies.

Pleased by her offer, Krishna thought of straightening the charming-faced Kubjā, who had a three-fold hump, by physiotherapy technique. The relevant shlokas from the two books describing the technique are given below -

पद्भ्यामाक्रम्य प्रपदे द्व्यंगुल्युत्ताणपाणिना।
प्रगृह्य चुबुकेऽध्यात्ममुदनीनमदच्युतः।। [भा० १०।४२]
तां च कुब्जां स्थगोर्मध्ये द्व्यंगुलेनाग्रपाणिना।
शनैः सम्पीडयामास कृष्णो लीलाविधानवित्।। [ह० २।२७]
सा तदर्जुसमानांगी बृहच्छ्रोणिपयोधरा।
मुकुन्दस्पर्शनात् सद्यो बभूव प्रमदोत्तमा।। [भा० १०।४२]
साच मग्नं स्थगुं मत्वा स्वायतांगी शुचिस्मिता।
जहासोच्चैः स्तनतटी ऋजुयष्टिर्लता यथा।।
प्रणयाच्चापि कृष्णं सा बभाषे मत्तकाशिनी।
क्व यास्यसि मया रुद्धः कान्त तिष्ठ गृहाण माम्।।
तौ जातहासावन्योन्यं सतलाक्षेपमव्ययौ।
वीक्षमाणौ प्रहसितौ कुब्जायाः श्रुतविस्तरौ।। [ह० २।२७]

[Ref. Bhāgvata Purāna, Skandha 10, Chapter 42 and Harivansha Purāna, part 2, Chapter 27]

Meaning: Pressing her toes with his feet, and holding her chin with two fingers of his hand placed under it, Lord Krishna pulled it upward. And, being well-versed in the art of physiotherapy, he slowly pressed at the middle of her hump, with a two finger-portion of the front part of his hand.

Then, by the touch of Krishna, she instantly became a beautiful woman, with a straight body, equitable organs, and robust breasts and hips. Knowing that her hump had been removed, she, with a well-stretched body, a charming smile and protruded breasts, looking like a creeper on a straight stalk, laughed loudly. Apparently intoxicated with joy, she said to Krishna affectionately, ``O beloved, withheld by me, where will you go? Stay here. Accept me." Hearing Kubjā's solicitation, Balarāma and Krishna had a hearty laugh, looking at each other striking their palms of hands.

From the narration, it is clear that Krishna was well-versed in the art of physiotherapy. He used the technique of physiotherapy, not magic, to treat and remove the hump of the young girl, Kubjā. It also shows the superb character of the two young boys, Krishna and Balarama.

CHAPTER 11

Vedic Mathematics

In the book *Vedic Mathematics* by the Shankaracharya of Kanchi, 16 Sutras (सूत्र) and 13 Sub-sutras (उप–सूत्र) from ancient Sanskrit literature, have been given. With the help of these Sutras and Sub-sutras very quick calculations to solve mathematics problems can be performed. The ancient Indian scientists and Astronomers used these and other Sutras to solve very intricate differential equations etc., very quickly.

One of the Sutras is "ऊर्ध्वतिर्यग्भ्याम्". By the application of this Sutra, multiplication of very large numbers can be done in a single line in a very short time. Upto a few years ago, the algorithm for multiplications in computers was based on the usual school method of multiplication but the latest algorithm is based on the method suggested in the above Vedic Sutra, thus saving a lot of computer time.

Another Sutra is "एकाधिकेन पूर्वेण". By the use of this Sutra, the process of division is simplified to a large extent and the process can be completed in much shorter time.

Indian mathematicians had calculated the value of π upto 31 places of decimal and coded it in a Sanskrit shloka for easy memorization. The shloka reads.

गोपीभाग्य मधुव्रात श्रृंगिशोदधिसन्धिंग ।
खलजीवितखाताव गलहालारसंधर ।।

Giving the value of π as

3.1415926535 89793 2384026433832792

accurate upto all the 31decimal places.

The "Pythagoras" theorem is clearly enunciated in the Shulva Sutra (शुल्व सूत्र), centuries before the birth of "Pythagoras". In contrast to the very lengthy proof of this theorem, as taught in the schools, the Indian Mathematicians had given five very short and simple proofs of this theorem, as illustrated in the book "Vedic Mathematics".

The concept of zero and infinity was introduced by ancient Indian mathematicians. In Yajurveda, in a 'Mantra', counting in decinary system has been recorded up to billions and more. This 'Mantra' was written by sage Medhatithi, in about 7900 BCE, when the rest of the world, did not know counting, beyond a thousand or so.

PART-III

Flights of Fancy for Futuristic Fundas

CHAPTER 12

Aeroplanes

In the ancient Sanskrit book *Brihadvimãna Shastra* design of aeroplanes, including metallurgy for manufacturing lightweight, high strength, low conductivity and fire-proof alloys, as well as instruments to be installed in the aeroplanes has been described. Four categories of aeroplanes, with more than 100 sub-categories have been mentioned. This book, in turn, is a compilation from more than 100 texts, which are given as references in various chapters of the book. Some of these reference books, must be containing lots of advanced science and technology. All-out effort should be made to locate and decipher these books.

Such vast literature, relating to Science and Technology with reference to design of aeroplanes, indicates that aeroplanes must have been manufactured in ancient India. Reference to the use of aeroplanes is also available in the ancient Indian history books.

In addition to the Pushpak Viman of *Ramayana*, there is a reference in Harivansha Purana, that Kartavirya Arjuna (about 8200 BCE) had a 400-metre-long aeroplane, which was propelled by solar energy. The relevant shloka is given below:

यस्तु बाहु सहस्रेण सप्तद्वीपेश्वरोऽभवत् ।
जिगाय पृथिवीमेको रथेनादित्यवर्चसा ।।

which literally means:

"He (Kartavirya Arjuna) became the emperor of the seven continents, after conquering the Earth with his thousand Bahu (Bahu is a unit of length, approximately equal to 40 cm) long plane, propelled by solar energy".

Unfortunately, this shloka has been, grossly mistranslated. The word 'बाहु–सहस्र' has been misinterpreted as meaning that he had a thousand arms. How can it be imagined that a person could have 500 arms emanating from each shoulder?

Actually, here the word 'बाहु–सहस्र' is evidently an adjective of 'रथ' (plane), as both these words have a third विभक्ति as per the rules of Sanskrit grammar. Thus, 'बाहु–सहस्र' as an adjective means "a thousand 'बाहु' or 400 metres long". Similarly, the word 'आदित्यवर्चस्' also has third विभक्ति and is, therefore, an adjective of 'रथ'. As 'वर्चस्' means energy, therefore, 'आदित्यवर्चस्' means having (or propelled by) solar energy.

In Mahãbhãrata era also, king Shishupãl's friend Shãlva, had a versatile plane named "Saubh", with which he attacked Dwaraka. The relevant shlokas in *Bhãgvata Purãana*, read:

स लब्ध्वा कामगं यानं तमोधाम दुरासदम्।
ययौ द्वारवतीं शाल्वो वैरं वृष्णिकृतं स्मरन्।।
बहुरूपैकरूपं तद् दृश्यते न च दृश्यते।
मायामयं मयकृतं दुर्विभाव्यं परैरभूत्।।
क्वचिद् भूमौ क्वचिद् व्योम्नि गिरिमूर्ध्नि जले क्वचित्।
अलातचक्रवद् भ्राम्यत् सौभं तद् दुरवस्थितम्।।

[भा. 20/96]

Meaning: "Taking a plane, which moved as per the will, was an abode of darkness (as no light or rays were reflected from it, and, therefore, did not come under surveillance of radar), was difficult to be attacked, Shalva proceeded to Dwaraka, remembering his enmity with Vrishnis.

That plane sometimes appeared as many, sometimes as one and sometimes became invisible. It was elusive, was designed by 'मय'. (There were two schools of Engineering and Technology, in ancient India – one established by Vishvakarmã and the other established by मय). It was difficult to be located by the enemies. Sometimes it was on the land, sometimes in the sky, sometimes at the top of the mountain, sometimes on sea-surface. The 'Saubh' moved like a hurricane and could not be pin-pointed".

From the above description, it can be observed how versatile and technically advanced this plane was. Evidently, highly advanced aeroplanes were there in Rãmayãn as well as Mahãbhãrata era.

CHAPTER 13

Space Travel

In *Harivansha Purãana,* which is considered to be a separate part of Mahãbhãrata (महाभारत खिल–भाग), there is a mention of space-travel by Raivata Kukudmi, grandson of Manu's grandson Anarta. [Ref: Book 1, Chapter 10]. It is written that Raivata, along with his daughter, Suvratã Revati, went to attend a music concert, near Brahma-Loka and when he came back to his city Kushasthali from where he had started for the space-travel, he saw that Kushasthali had been renovated and occupied by Yãdavas and renamed as Dvãravati (Dvãraka). The relevant shlokas are quoted below:

आनर्तस्य तु दायादो रेवो नाम महाद्युतिः।
आनर्तविषयस्त्वासीत् पुरी चास्य कुशस्थली।।
रेवस्य रैवतः पुत्रः कुकुद्मी नाम धार्मिकः।

स कन्यासहितः श्रुत्वा गान्धर्वं ब्रह्मणोऽन्तिके।।
आजगाम युवैवाथ स्वां पुरीं यादवैर्वृताम्।
कृता द्वारवतीं नाम्ना बहुद्वारां मनोरमाम्।।
ततः स रैवतो ज्ञात्वा यथातत्त्वमरिंदम।
कन्यां तां बलदेवाय सुव्रतां नाम रेवतीम्।।
दत्त्वा जगाम शिखरं मेरोस्तपसि संस्थितः।

Meaning: Ãnarta's son was Reva. Ãnarta's kingdom was confined to the city Kushasthali. Reva's son was Raivata Kukudmi. He returned after attending a music concert near BrahmaLoka, to his city, as young as he was when he started off for the space travel. His city Kushasthali had been occupied by Yãdavas and renamed as Dvãravati and renovated as a charming city with many gates. Then, after knowing the factual position (taking into consideration the time-warp due to speed of travel nearly equal to that of light), he gave his daughter named Suvrata Revati to Balarãma (in marriage) and himself went to Meru's top for leading an ascetic life.

As per the description of the Surya dynasty in which the names of Manu's descendants have been given, about 90 kings had ruled from Manu upto Brihadbala, who was killed in Mahãbhãrata war by Abhimanyu. Thus, about 86 generations had elapsed during the period of the space travel by Raivata Kukudmi and his daughter Suvratã. Clearly, it took them several thousand years to perform the to-and-fro journey.

In Mahãbhãrata [Parva 1, Chapter 125], it is mentioned that when the sages from the forest came to Hastinãpura to hand over the Pãandavas and Kunti, alongwith the bodies of Pãndu and Madri, to Bhishma and Dhritarãshtra, the sages after handing them over, disappeared along with the mystics in a moment in full view of all the Kurus, indicating that the mystics could transport themselves taking along with them other persons, with the speed of light, or instantaneously. The relevant shlokas are:

इमे तयोः शरीरे द्वे पुत्राश्चैते तयोर्वराः।
क्रियाभिरनुगृह्यन्तां सहमात्रा परन्तपाः।।
एवमुक्त्वा कुरून्सर्वान् कुरूणामेव पश्चताम्।
क्षणेनान्तर्हिताः सर्वे तापसा गुह्यकैः सह।।

Meaning: "These are their two bodies and these are their worthy sons. Receive them along with their mother with due respect". So saying to the Kurus, the ascetics, along with the mystics, instantaneously, disappeared, in full view of the Kurus.

In modern science, they have so far been able to teleport only sub-atomic particles, at the speed of light, from one place to another. In ancient India, probably they could teleport persons from one place to another, with speed of light, or instantaneously. If ancient scientific literature is searched, may be we can find a clue for this advanced research.

As the space travel by king Raivata and her daughter Revati, took about 6,000 years for to and-fro journey, there is a possibility of some civilisation on a planet at a distance of about 3,000 light years from the Earth, towards the centre of our galaxy, as the word 'ब्रह्मान्तिके' indicates that they had gone on space-travel, towards the centre of our galaxy.

It would be worthwhile, to search for another civilisation, at the location indicated.

Arjuna's Space Travel

Another event of space travel is mentioned in Mahabharata by Rishi Vyasa, wherein it is stated that during the period of 12 years of exile of Pandavas, Arjuna went to Indraloka, stayed there for five years to get training in super-weapons.

He then returned to the Earth equipped with the super-weapons acquired from Indra.

The relevant Shlokas are given below -

इन्द्रे ह्यस्त्राणि दिव्यानि समस्तानि धनञ्जय।
वृत्राद्भीतैस्तदा देवैर्बलमिन्द्रे समर्पितम्।।

Arjun, the gods, afraid of Vritra, had, at that time, placed all the Divine weapons in the custody of Indra.

तान्येकस्थानि सर्वाणि ततस्त्वं प्रतिपत्स्यसे।
शक्रमेव प्रपद्यस्व स तेऽस्त्राणि प्रदास्यति।।

You will get all of them collected at one place. Go to the shelter of Indra. He will provide you the weapons."

निदेशाद्धर्मराजस्य द्रष्टुं देवं पुरंदरम्।
धनुर्गाण्डीवमादाय तथाक्षय्यौ महेषुधी।।

With Yudhishthir's directive, Arjun started off, desirous to see Indra, taking his bow.

हिमवन्तमतिक्रम्य गन्धमादनमेव च।
अत्यक्रामत्स दुर्गाणि दिवारात्रमतन्द्रितः।।

Crossing the Himalayas and Gandhmadan, he passed through difficult terrains, travelling day and night, without any lassitude.

इन्द्रकीलं समासाद्य ततोऽतिष्ठद्धनञ्जयः।
अन्तरिक्षे हि शुश्राव तिष्ठेति स वचस्तदा।। (म० ३/३७)

Reaching Indra-keel, Arjuna stopped. Then he heard a voice in space, saying, "Stop."

ततश्चिन्तयमानस्य गुडाकेशस्य धीमतः।
रथो मातलिसंयुक्त आजगाम महाप्रभः।।

As Arjun was pondering over it, a very brilliant chariot, with Matali (Indra's chariot driver), came there.

ततोऽर्जुनो हृष्टमना गंगायामाप्लुतः शुचिः।
आरुरोह रथं दिव्यं द्योतयन्निव भास्करः।।

Then Arjun, with glad mind, purified by taking a dip in Ganga, mounted the divine chariot glowing like the Sun.

स तेनादित्यरूपेण दिव्येनाद्‌भुतकर्मणा।
ऊर्ध्वं प्रचक्रमे धीमान् प्रहष्टः कुरुनन्दनः।।

The joyful Arjun moved upwards by that wonderful chariot looking like the Sun.

ददर्शाद्‌भुतरूपाणि विमानानि सहस्रशः।
न तत्र सूर्यः सोमो वा द्योतते न च पावकः।। (म० ३/४२)

He saw thousands of wonderful planes. There was neither the Sun nor the Moon shining nor Fire.

नन्दनं च वनं दिव्यमप्सरोगणसेवितम्।
ददर्श दिव्यकुसुमैराह्वयद्भिरिव द्रुमैः।।

And saw the divine Nandan forest inhabited by nymphs and enchanting trees with divine flowers.

स तद्दिव्यं वनं पश्यन्दिव्यगीतानिनादितम्।
प्रविवेश महाबाहुः शक्रस्य दयितां पुरीम्।।

Seeing that divine forest resounding with divine songs, he entered Indra's beloved city.

ततः पार्थो महाबाहुरवतीर्य रथोत्तमात्।
ददर्श साक्षाद्देवेशं पितरं पाकशासनम्।।

Then, descending from the chariot, Arjun saw the father Indra, Lord of gods, before his eyes.

ततः शक्रासने पुण्ये देवर्षिगणसेविते।
शक्रः पाणौ गृहीत्वैनमुपावेशयदन्तिके।। (म० ३/४३)

Then Indra, holding him by hand, made him sit beside him, on the auspicious Indra-throne, attended on by god Rishis.

एवं सम्पूजितो जिष्णुरुवास भवने पितुः।
उपशिक्षन् महास्त्राणि ससंहाराणि पाण्डवः।। (म० ३/४४)

Thus honoured, Arjun stayed in the father's house, getting training for great weapons, along with the method of neutralizing them.

उषित्वा पंचवर्षाणि सहस्राक्षनिवेशने।
अवाप्य दिव्यान्यस्त्राणि सर्वाणि विबुधेश्वरात्।।
आग्नेयं वारुणं सौम्यं वायव्यमथ वैष्णवम्।
ऐन्द्रं पाशुपतं ब्राह्मं पारमेष्ठ्यं प्रजापतेः।।
यमस्य धातुः सवितुस्त्वष्टुर्वैश्रवणस्य च।
तानि प्राप्य सहस्राक्षादभिवाद्य शतक्रतुम्।।
अनुज्ञातस्तदा तेन कृत्वा चापि प्रदक्षिणम्।
आगच्छदर्जुनः प्रीतः प्रहृष्टो गन्धमादनम्।। (म० ३/१६४)

Staying in Indra's house for five years, and obtaining from the Lord of gods, all the divine weapons - Agneya, Varun, Saumya, Vayavya, Aindra, Pashupat, Brahma, Parameshthya, Prajapatya, Yam's, Dhata's, Surya's, Tvashta's, Vaishravan's - getting these and bidding adieu to Indra, and going around him keeping him on the right, the pleased Arjun joyfully came to Gandh-madan.

It is clear from the above narration that for Arjun's journey to Indra Loka, Indra sent his spaceship which landed at a space station named Indra-keel, located in the high ranges of Himalayas on the bank of River Ganga.

CHAPTER 14

Electronic and Other Gadgets

In *Brihadvimãna Shastra,* several gadgets are mentioned, names of some of which are given below:

1) *Guhã garbhãdarsha Yantra* (गुहा–गर्भादर्श यन्त्र)- As the name indicates, it was a device by which any concealed (or underground) objects could be located, as by ultrasonic or magnetic response imagery in modern science.
2) *Vishvakriyadarsha Yantra* (विश्वक्रियादर्श यन्त्र)- It was probably a sophisticated type of television as the name indicates.
3) *Shabdãkarshana Yantra* (शब्दाकर्षण यन्त्र)- Probably, a sophisticated type of radio.

4) *Apasmãra Dhumra Prasãrana Yatra* (अपस्मार धूम्र प्रसारण यंत्र)- A device from which a gas was spread, which could result in the loss of memory of the enemies.

5) *Stambhana Yatra* (स्तंभन यन्त्र)-A device from which such waves were emitted which would paralyse the enemies.

Such sophisticted devices were available in the Mahãbharata era, is indicated from the following shlokas:

Telephones-

निर्ययुर्नगराच्छूराः कौरवाः पांडवैः सह।
उद्यानवनमासाद्य विसृज्य च महाजनम्।।
विशन्ति स्म तदा वीराः सिंहा इव गिरेर्गुहाम्।
उपस्थानगृहैः शुभ्रैर्वल्लभीभिश्च शोभितम्।।
गवाक्षैस्तथा जालैर्यन्त्रैः सांचारिकैरपि।
दीर्घिकाभिश्च पूर्णाभिस्तथा पद्माकरैरपि।।
[पर्व 1] Chapter 127]

Meaning:

The Kauravas along with the Pãndavas went out from the city (Hastinãpura) by big chariots and indigenous elephants. Leaving the greater Hastinapura and reaching the garden - forest, decorated with white rest-houses, with pinnacles, ventilators, nets and Communication instruments, oblong ponds filled with water as also lotus ponds, they entered it as lions enter the cave of the mountain".

The reference to context of these lines is when Duryodhan hatched a conspiracy to kill Bhima, he got an elaborate temporary water sports complex (उदक क्रीडन) constructed at the bank of Gangã, with sophisticated temporary rest-houses housing all facilities, including commucation instruments.

Robotic Alligater:

Another reference is regarding a robotic alligator contrived by Drona to test his disciples [पर्व 1] Chapter 132] The story is like this: Drona wanted to test his disciples. He got a contrived (परिकल्पित) alligator, placed in the Gangã and went to the Gangã, along with his disciples to take a bath. As he entered the Gangã, the contrived (robotic) alligator, impelled by time , caught hold of Drona's leg. Drona cried for help. Whereas the other disciples were non-plussed, Arjun, immediately shot the alligator with sharp arrows and the contrived alligator was shattered into many pieces. The relevant shloka reads:

स पार्थबाणैर्बहुधा खंडशः परिकल्पितः।
ग्रहः पंचत्वमापेदे जंघां त्यक्त्वा महात्मनः।।

Meaning: "That contrived (robotic) alligator broke variously into the five elements, piece by piece, by Arjuna's arrows, releasing Drona's leg".

CHAPTER 15

Super Weapons

In ancient India, super weapons were available right from the Surya dynasty king Sagar (about 6800 BCE), who used Agneya Astra (आग्नेय अस्त्र) to defeat his enemies, including Scythians and Greeks, who had earlier vanquished his father, king Bãhu, upto the Mahãbharata era (about 3100 BCE). Although the names of several super weapons are mentioned in Rãmãyana and Mahãbhãrata, their details are not available, except a vivid narration of the working of Aindra-Astra (Lightning weapon), used by Arjuna in his solo battle against the combined Kaurava Army when they attacked Virata city. The relevant shlokas are cited below:

ततः प्रहस्य बीभत्सुर्दिव्यमैन्द्रं महारथः।
अस्त्रमादित्य संकाशं गांडीवे समयोजयत्।।

यथा बलाहके विद्युत् पावको वा शिलोच्चये।
तथा गांडीवमभवदिन्द्रायुधमिवानतम्।।
यथा वर्षति पर्जन्ये विद्युद् विभ्राजते दिवि।
द्योतयन्ती दिशः सर्वाः पृथिवीं च समन्ततः।।
तथा दश दिशाः सर्वाः पतद्गांडीवमावृणोत्।
नागाश्च रथिनः सर्वे मुमुहुस्तत्र भारत।।
सर्वे शान्तिपरा योधाः स्वचित्तानि न लेभिरे।
संग्रामे विमुखाः सर्वे योधास्ते हतचेतसः।।
[पर्व 4, अध्याय 63]

Meaning: "Then Arjuna laughed and installed the divine Indra - weapon, like the Sun on the Gãndiv bow. Like the lightning in a thunder - cloud or fire on a hill-top, Gãndiv became curved like a rainbow. As lightning flashes in the sky, during rainfall, lighting up all the directions and the Earth all around, similarly Gãndiv, controlling all the ten directions, covered them all. All the elephants and charioteers there became unconscious. All the wariors became tranquil and lost their consciousness; all those wariors with their psyche benumbed, became averse to war".

From the above narration, it is evident that the electromagnetic waves, emanating from the lightning weapon, rendered the entire Kaurava army, unconscious.

Obviously, this weapon used electromagnetic waves, but these waves were of entirely different frequencies than those used in Micro-wave weapon.

In Valmiki Ramayana, names of more than hundred super weapons have been mentioned (Ref Bala Kanda Chapter 27, 28).

These weapons were developed in the two organizations, named Jaya (जया) and Suprabha (सुप्रभा), established by king Daksha (दक्ष)) (ref : Valmiki Ramayana Bala Kanda, Chapter 21) -

सर्वास्त्राणि कृशाश्वस्य पुत्राः परम धार्मिकाः।
कौशिकाय पुरा दत्ता यदा राज्यं प्रशासति।। १–२१–१३

All the super-weapons - the sons of Krishashva, were given to Adi-Vishvamitra when he was ruling over a kingdom. [1-21-13]

तेऽपि पुत्राः कृशाश्वस्य प्रजापतिसुतासुताः।
नैकरूपा महावीर्या दीप्तिमंतो जयावहाः।। १–२१–१४

These super-weapons - the sons of Krishashva, are the sons of daughters of Daksha. They are of various forms, and are very powerful, dazzling, and victorious. [1-21-14]

जया च सुप्रभा चैव दक्षकन्ये सुमध्यमे।
ते सुवातेऽस्त्रशस्त्राणि शतं परमभास्वरम्।। १–२१–१५

Jaya and Suprabha, the delicate daughters of Daksha, produced a hundred extremely resplendent weapons and super-weapons. [1-21-15]

पंचाशतं सुतान् लेभे जया लब्धवरा वरान्।
वधायासुरसैन्यानामप्रमेयानरूपिणः।। १–२१–१६

Jaya begot fifty invincible, unfathomable, great sons, for annihilation of the Asura armies. [1-21-16]

सुप्रभाजनयत् चापि पुत्रान् पंचाशतं पुनः।
संहारान्नाम दुर्धर्षान् दुराक्रामान् बलीयसः।। १–२१–१७

Then Suprabha, begot fifty unassailable, unconquerable anti-weapon weapons. [1-21-17]

These verses are narrated by sage Vasishtha to Dashratha to assure him that prince Rama would be safe in the hands of sage Vishvamitra, as he possessed formidable super-weapons, inherited by him from Adi-Vishvamitra, to whom they were given by Krishashva, when Adi-Vishvamitra was ruling over a small kingdom.

It may be noted here that in *Bhagavat Puarana* (Skandha 9, chapter 15) and *Harivansha Purana* (Part 1, chapter 27) Adi-Vishvamitra was the 12th descendent of Vijaya - a son of Pururava, who was Manu's daughter, Ila's son. Thus, Adi-Vishvamitra was 14th descendent of Manu's daughter Ila. While Manu's central kingdom, with Ayodhya as its Capital, was inherited by Ikshavaku, a portion of the kingdom with Pratishthanpur as its Capital was inherited by Ila (who had become Sudyumna after a sex-change operation.) While the kingdom of Pritishthanpur was inherited by Aayu, the eldest son of Pururava, a small portion of the kingdom was inherited by Vijaya, which was subsequently inherited by Adi-Vishvamitra.

Thus, Adi-Vishvamitra was the 15th descendent of Manu, through his daughter Ila, and Samhatashva was the 15th descendent of Manu through his son Ikshvaku. Therefore, Adi-Vishvamitra would be a contemporary of Samhatashva and his son Krishashva in around 8300 BCE. Adi-Vishvamitra became a Sanyasi (Rishi or researcher) and abdicated his small kingdom. Probably, he may have handed over his kingdom to Krishashva in exchange for the two super-weapon producing organisations/labs - Jaya and Suprabha.

While scientists of Jaya developed various kinds of super weapons, those at Suprabha developed weapons for the annihilation of the super-weapons, just like anti-missile systems are used to annihilate ballistic missiles. Krishashva, a son of Samhatashva, got these two organizations as inheritance from his father; while the kingdom was inherited by his sister Harimavati. (ref: Harivansh Purana, Part 1, Chapters 11,13) The Adi-Vishvamitra established the disciple tradition (शिष्य परम्परा) of Vishvamitras, just like later Adi-Shankracharya established the disciple tradition of Shankaracharyas. In this manner, the two super-weapon labs were later inherited by another Vishvamitra, a contemporary of Dashratha.

It may be mentioned here that the Vishvamitra Tradition was not the only School of sciences in ancient Bharat. There was a Vasishtha School, established by Adi-Vasishtha in around 9300 BCE. While the Vishvamitra school specialized in super-weapons, the Vasishtha school is indicated to be a school dedicated to Robotics. Valmiki Ganita (गाणित) School did research in Fundamentals of Physics, Math and Computing. Also, there was Vishvakarma (विश्वकर्मा) school of Engineering and

Technology and also a Maya (मय) school of Engineering Technology and Astrophysics.

In this context, it may be observed that the Rishis of ancient Bharat were, actually, researchers. Now a question arises that if Vishvamitra had with him such advanced super-weapons, why could he have not used them against the vandals of Ravana, instead of asking prince Rama to perform the task for him.

The answer is simple if we compare it to a similar present-day scenario. While the scientists of DRDO and ISRO develop the weapons, they are not the ones using them, instead they hand them over to the Army officers for use against the enemy. In the same manner Vishvamitra had to hand over the super-weapons he possessed to a capable king or prince. If we read the scriptures in proper perspective, without any preconceived prejudices in our mind, these facts become crystal clear.

In Valmiki Ramayana, the above mentioned super weapons are mentioned as the sons of Krishashva. Also, Jaya and Suprabha are mentioned as the two daughters of king Daksha. It cannot be plausibly or logically imagined that the two girls can give birth to super-weapons. The only logical inference is that Jaya and Suprabha were super-weapon making organizations/labs/ industries which king Daksha had established. Even in present day parlance, when an intellectual conceptualizes a new project, he is claimed to be its "Father". We must, therefore, apply logic and deduce that Jaya and Suprabha were indeed such projects that were conceptualized by king Daksha, and dear to his heart. Further, the weapons developed in these two labs are mentioned as the sons of these two daughters of king Daksha. As they were at some time owned by Krishashva, the weapons, thus, became his sons too. These logical interpretations also clear up the anomaly of how could Krishashva be married to the daughters of king Daksha as historically Daksha lived several thousand years before Krishashva. It may be noted that it is nowhere mentioned that Jaya and Suprabha were married to Krishashva. The relevant verses only state that the super-weapons were their sons, and these were also sons of Krishashva.

This interpretation could be made by applying just one layer of straight forward logic borrowed from the modern day parlance. But there also exist certain idioms/encryptions that were in use during ancient times which are no longer a part of our language and it is, therefore, even more difficult to interpret them in their proper sense. In those cases

1. We have to look for context clues and connotations to properly understand the otherwise fantastical episodes of our ancestors' history.
2. Another prerequisite is to firmly believe that at no time in recent or ancient history, anywhere in the whole big universe, were the eternal principles of physical sciences ever flouted. So whatever seems impossible today, was most probably impossible in that era too.
3. We must also sift carefully through the literature to separate folk-lore from true historical facts. For example, we know that both Valmiki Ramayana and Mahabharata are essentially historical accounts of their respective eras, but they also have some interpolations/extrapolations that need to be discarded before their core story becomes clear.

4. A thorough and complete knowledge of Sanskrit is a must to even start on a journey down this road. Because most of the translations of these historical scriptures that we find today are by European authors. Their foreign minds cannot be expected to discern the true meanings of a language that is not native to them. Sanskrit revels in synonyms, and many words have multiple synonyms, which are sometimes instantly clear to us as they continue to be used in our present-day Sanskrit-based regional languages as well as Hindi. Even among the local scholars, we can find only a few who have the patience to dive into dictionaries to find dual meanings or synonyms. Most modern-day translators do not even correctly apply the Vibhakti and Dhatu rules in their hurry.
5. Another important aspect is to shed the prejudice born from preconceived notions. When we read, we must read as though it is the first time we are hearing of the event.
6. Wipe away all the temptations of attributing miraculous deeds to our ideals

(Rama, Hanumana, Krishna etc). Remember that even when God Himself is telling about His conduct as an Avatar, He clearly says that, when in human form, He follows all its physical constraints. He works very hard and becomes the best human that there could be in those circumstances. His conduct is impeccable, His heart pious, His intellect sharp. But, at the end of the day, He is Human. And, therefore, it is important that we discard all ideas of miracles either as hyperbole or encrypted/idiomatic use of language by the narrator.

Let's now look at a few such episodes, which mention names of real natural entities that exist to this date, along with names of Rishis (scientists) that lived in that era.

1. It is a popular story connected with the story of Descent of Ganga, that after Ganga started to flow through the foothills of Himalayas and down in the great northern plain, she irked the sensibilities of Rishi Jahnu. It is told that he filled up all the water of Ganga in the hollowed-out palms of his hands and drank it up.

2. Connected to the same story, we come across the story of Rishi Agastaya being approached by the Devas to help locate the Danavas hiding inside the waters of the ocean. It is told that Rishi Agastya filled the ocean in the hollowed palms of his hands and drank it up.
3. To a logical mind, both of these stories are nothing but humbug. It is indeed impossible to imagine that Rishis (regular human beings where their physical capabilities were concerned) were capable of such miracles. But if we now shed our prejudice that tells us that Rishis were essentially men of only spiritual knowledge who lived in deep forests for the sole purpose of mingling their souls with the supersoul. Instead, now we must proceed with the essential assumption that each of the Rishis was a highly accomplished scientist who often specialized in more than one field. For example, we know that Rishi Valmiki was a historian and a poet, but many of us do not know that he ran a school (university) of math and physics. Similarly, Rishi Vyas was a gynaecologist

par excellence (he was involved in the birth of Dhritarashtra, Pandu and Vidura; after that in the birth of the hundred Kauravas too. But he was also a poet and a historian.

4. So when we come across multiple references of similarly illogical events, we must then look for contextual clues and decipher their true meaning.
5. So we now search for another reference of the same phrase - drinking by filling up in the hollowed palms of one's hands (अंजलिभिः पातुं). In Valmiki Ramayana itself (Kishkindha Kanda, chapter 28, verse 8), we come across this verse that says -

मेघोदरविनिर्मुक्ताः कर्पूरदलशीतलाः ।
शक्या अञ्जलिभिः पातुं वाताः केतकगन्धिनः ।।

Meaning: "The winds released from the clouds, cold as camphor, and smelling like the Ketaka flower, are (so pure that they are) capable of being drunk up in the hollowed palms of hands."

This is the description narrated by Shri Ram to Lakshman describing the rainy season and its various elements while residing in a cave near Kishikindha.

When we apply this meaning of the idiom drunk up in the hollowed palm of the hands to the previous two instances, their true meaning becomes very much clear, that when Ganga initially descended from the slopes of Himalayas and entered the plains, her water became unclean, and, therefore, unfit for consumption by humans. Noticing this, Rishi Jahnu then made efforts as a scientist to purify it, so that it could be filled in the palms of hands to drink.

Similarly, Rishi Agastaya may have set up a desalination plant near the ocean. Since the ocean water is far from clean and quite impotable; the Rishi scientist in this manner may have fulfilled the potable water requirements of the people living in the area, by making the ocean water fit for filling in the hollowed out palms of the hands and drunk.

CHAPTER 16

Heavy Machinery

When the causeway (सेतु) between India and Lanka was constructed, heavy earth-moving machinery was used to excavate huge rocks from the mountains and transport them to the sea-shore. It is clear from the following shloka in the Valmiki Ramayana.

हस्तिमात्रान् माहाकायाः पाषाणांश्च महाबलाः ।
पर्वतांश्च समुत्पाट्य यन्त्रैः परिवहन्ति च ।।
[युद्ध–कांड, Chapter 22]

Meaning: The mighty and hefty (Vãnavas) transported huge stones of the size of elephants, after excavating them from the mountains with machinery".

It is clear that they had such heavy rock-cutting machinery, which could cut out huge stones of the size of elephants. Such heavy rock cutting machinery is not available even in modern-day technology.

It may be noted that the causeway was constructed in just five days. The amount of rock needed for the causeway was millions of cubic metres. Use of heavy rock-cutting machinery was, therefore, essential to complete the job in time. The time available for completing the job was limited, because Rãavana had given Sita two-months ultimatum, after which his (Rãvana's) cooks would cut her body into pieces, cook them and serve the cooked fragments to him for satisfying his lust for Sita. The relevant shloka reads:

द्वाभ्यामूर्ध्वं तु मासाभ्यां भर्तारं मामनिच्छतीम् ।
मम त्वां प्रातराशार्थे सूदाश्छेत्स्यन्ति खंडशः ।।
[सुंदर कांड, Chapter 22]

Meaning: "If you do not accept me as your husband within two months, then after that, my cooks will cut you in fragments for (preparing) my breakfast".

As already some weeks had elapsed since Rãvana's ultimatum of two months, the time left was really limited.

In Mahãbhãrata era also, there is an indication of use of heavy earth-moving machinery during excavation of Govardhana mountain, planned by Krishna to create a safe sanctuary for safety of cowherds and cows, from the fury of wind and rains.

The description of the scenario, when the mountain was being excavated, as given in *Harivansha Purana* [Book 2, Chapter 18] is a replica of the scenario created when the abutments of a large dam in a mountainous region, are being excavated by heavy machinery as can be vouchsafed by those who have had the opportunity of witnessing such activity. The relevant shlokas are quoted below:

भूमेरुत्पाट्यमानस्य तस्य शैलस्य सानुषु।
शिलाः प्रशिथिलाश्चेलुर्विनिष्पेतुश्च पादपाः।।
शिखरैर्घूर्णमानैश्च सीदमानैश्च पादपैः।
विधृतैश्चोच्छ्रितैः श्रृंगैरगमः खगमोऽभवत्।।
कानिचिच्छिथिलानीव संच्छिन्नार्द्धानि कानिचित्।
गिरेर्मेघप्रविष्टानि तस्य श्रृंगानि चाभवन्।।
विषमैश्च समीभूतैः समैश्चात्यन्त दुर्गमैः।
व्यावृत्तदेहः स गिरिरन्य एवोपलक्ष्यते।।
निवेश्य तं करे शैलं तोलयित्वा च सस्मितम्।
प्रोवाच गोप्ता गोपानां प्रजापतिरिव स्थितः।।

एतद्देवैरसंभाव्यं दिव्येन विधिना मया।
कृतं गिरिगृहं गोपाः निर्वातं शरणं गवाम्।।

Meaning: "As the mountain was being excavated from the ground, stones at the ridge of the mountain, loosened and started falling and trees gushed out crumbling. With the tops being shaken, trees sinking down and high peaks trembling, the sky-scraper mountain became inaccessible. Some of the peaks of the mountain piercing the clouds, became as if unsteady and some were half-broken. With its uneven surfaces having become even and the even ones having become inaccessible and its body ripped apart, the mountain appeared to be altogether different. Bringing the mountain under his control and making an assessment, Krishna the protector of cowherdsmen, stood like Brahmã and said with a smile.

"Gopas this wind-proof mountain-abode, the shelter for the cows, impossible to be created even by gods, has been prepared by me by a divine method".

It may be noted that the word 'करे' in Sanskrit can be translated as "on the hand" as well as "in the hand". Thus, the clause निवेश्य तं करे शैलम्

could be translated as "Placing that mountain on his hand", or "Bringing that mountain in his hand i.e. under his control". Obviously, the latter meaning is the plausible meaning and not the former one, as is usually translated.

The scenario described above should be seen from a scientific viewpoint and not as a mythical miracle. From the scientific point of view, the Govardhan mountain was got excavated by Lord Krishna with heavy earth-moving machinery. The whole project was so planned that the mountain was ripped apart, creating a wind-proof shelter.

It is evident that heavy earth-moving machinery was available in the Ramayana era (around 5000 BCE), as well as in Mahābhārata era (around 3000 BCE) and also Bhagirath's era (around 6500 BCE) as he could not have made Ganga descend, without the help of heavy machinery.

CHAPTER 17

Descent of Bhagirathi (Ganga)

Our ancient scriptures are full of historical accounts that are a treasure trove of knowledge but only when approached with a scientific perspective. Reason being that these scriptures often assume a symbolic tone, which is only indicative of a highly developed style of writing.

While reading our Vedic texts, we must keep in mind that the civilization of our ancestors had at its centre such a developed form of science that is yet to be achieved by modern humans. It is but natural that this knowledge was an integral part of everything they did. The sages that we often assume to be old ascetics withdrawn into a solitary existence of spiritual austerities in the forests, were instead highly accomplished scientists and researchers, who led the various elite Universities scattered across the country in serene country-side settings.

When we read the texts written by our extremely accomplished ancestors in a mere spiritual context, we get tangled in the maze of their symbolic surface stories which then give rise to imaginative myths rooted in concepts of boons, curses, spells and magic. And the very same texts, when approached with a scientific intellect, reveal their hidden secrets to us, translating into tangible real accounts of their era; free from imaginative discrepancies.

There are hundreds of such stories—one of the most intriguing is that of the Descent of Ganga. This chapter delves deep into the verses that tell about it and strives to find their intended meaning.

The Pauranic Tale:

In the Puranas, the tale of Ganga is told in this manner—

In the famed clan of Raghu, there was a king named Sagar, who in his old age was blessed with 60,000 sons at once. Once this king organised an event for the betterment of his people. A horse, which was an important part of this ritual, was abducted by Indra, the King of gods, and hidden in the netherworld. Sagar sent his 60,000 sons to all corners of the Earth to look for this horse. They found it in the hermitage of sage Kapil. Assuming that the sage had stolen the horse, the princes started abusing the sage and interrupted his spiritual trance. The sage who had been sitting in meditation for years opened his eyes and looked at the 60,000 princes. Such was the power of his gaze that the unruly princes were reduced to ashes.

In the absence of last rites, the souls of these 60,000 princes were left roaming as ghosts and ghouls instead of finding a passage to the next realm. When a descendent of Sagar, Bhagirath, son of Dileep, heard about their misfortune, he took a vow to end their misery by bringing sacred Ganga to Earth to wash away their sins.

To fulfill his vow, first Bhagirath observed austerities to acquire the blessings of Brahma. When Brahma granted him a boon, he requested the Lord to ask the Sacred River to leave her heavenly abode and descend down to Earth.

Brahma ordered Ganga, accordingly. But she became extremely upset at the prospect of having to abandon her celestial station. She, in turn, threatened to fall to Earth with such abundant force that it would destroy everything in its path. Then, Bhagirath prayed to Shiva to help cut down her force.

Ganga in her pride started to flow down from heaven with unparalleled speed, but as she fell down she got caught in the locks of Lord Shiva's matted hair. She swirled around in that maze for long, unable to find a way out, then on Bhagirath's request, Shiva let her out in a gentle manner. Ganga had become even more sacred after being touched by Shiva in this process. From here, she flowed out in several streams in three directions. For this reason, she got the new name of "Tri-path-ga", meaning a river that flows on all three ways.

She also got the name of Bhagirathi, as she had descended on Earth only through the sustained efforts of Bhagirath. A Hindi phrase that took form from this incident was - "Bhagirath Prayatn" - applicable to intense efforts made by any person to reach their goal.

Skeptical Analysis of the Above Event

If one reads the story in this form, it appears to be imaginative and unreasonable in more ways than one. But that would be unthinkable for our elite ancestors who lived their lives in the most logical manner and always based everything on science and intellect alone, as is clear from the following verse from the *Bhagavata Purana* (Skandh 10, Chapter 77)

एवं वदन्ति राजर्षे ऋषयः केचनान्विताः।
यत्स्ववाचो विरुध्येत नूनं ते न स्मरन्त्युत।।

Meaning: "Some wise men say that anything (talk, book, principle, event) that does not match the principles of logic should definitely be discarded (as untrue)."

For 'magic' to be conjured or even its very existence finds no support from any of our scriptures. Only those who have an under-developed logic tend to believe in the concept of magic, as they cannot fathom the actual science or series of events leading to the apparently impossible 'deeds' in front of their eyes. And it is to take advantage of this lack of logical intellect, and to fool the common men further, that the ill-intentioned intellectuals further drag them down into the quagmire called 'magic'.

Contrary to that, those who walk in the footsteps of Yogeshvar Shri Krishna, they consider the very concept of 'magic' as false, because they know that the scientific principles that form the basis of this Creation, cannot be flouted even by the Creator Himself. Therefore, all stories and events that disprove the principles of science, have no basis in reality. But since we also know that Truth was paramount for our ancestors, that telling a lie was considered to be the biggest sin by them, we must approach their writings from another perspective—one that can make their logical sense. Only in this manner can we ever understand our real history.

When we re-read the same story and its scriptural sources with an analytical intellect, we see an entirely different form of it.

In this chapter, let us embark on a journey together to explore this unique interpretation - a historical account that is untouched by imagination or 'magical happenings', that instead confirms the fact that our ancient civilization was indeed much advanced in science and technology.

A Scientific Approach (Complete with ref and meaning of related verses)

When sage Vishvamira was traveling with Shriram and Lakshman towards the city of Mithila, upon reaching the banks of River Ganga, he told them the history of the great king Sagar's Yagya and the subsequent descent of Ganga. While this story sounds impossible and illogical on the surface, an intellectual quest by digging a bit below the surface brings out a very much possible narrative from the same verses; and becomes the reflective mirror for the science and engineering feats of the people of that era.

King Sagar was the 24th ancestor of Shri Ram. His father was emperor Bahu who ruled over the entire Aryavarta. When the Greeks and other foreign invaders got together to attack his kingdom, he was unable to put up a strong resistance and got defeated by them. After that, he retreated to the forest where his wife followed him. He soon died because of the extreme sorrow and depression overcame his will to live. His pregnant queen too was about to enter his burning pyre, when rishi Aurva came there and stopped her. He took her to his hermitage where her son, prince Sagar was born. Rishi Aurva looked after his education and training from the very beginning. When he came of age, he was taught the use of a weapon called the Agneya Astra (roughly translated - Fire Missile). (Ref - Harivansh Purana Part 1, Chapter 13-14)

With the help of this fierce weapon, Sagar was able to defeat the invading foreigners and regain his kingdom. In this manner, emperor Sagar ascended the throne of Aryavarta in the year 6797 BCE. (Ref - Capter 20). His son was Asmanjas and the son of Asmanjas was Anshuman. (Ref - Harivansh Puran, Part 1, Chapter 15) A long time after the birth of his grandson Anshuman, Emperor Sagar decided to perform a Yagna (Ref - Valmiki Ramayana, Balakand, Chapter 38).

Definition of the Word Yagna

In Chapter 4 of Bhagavad Gita Bhagwan Shri Krishna defines 13 different kinds of Yagnas out of which 12 are of the personal kind and relate to various techniques of Pranayama and self-discipline. And the 13th kind are called the material Yagnas which are, generally, for promoting welfare of the people around us and the society at large. The kings of ancient India often performed two of these material Yagnas in particular – the Rajsuya and the Ashvamedha.

Rajsuya was performed only by those rare kings who wanted to declare their supremacy over the other kings in the region. His Yagna would be a challenge to all others to either concede to him or fight against him. Only the kings with strong armies and clever diplomatic skills would aspire for such a title.

On the other hand, Ashvamedha could be undertaken by anyone, as its purpose was generally the welfare of masses. Many kings would, in fact, do one every year, by picking up a new welfare scheme to focus on. Over time, though the ritual of Ashvamedha got associated with many mythical and unreasonable opinions. Most probably, it was the interpreters and the translators who got stuck on only one most common meaning of the word 'Ashva' that is 'horse'. In Sanskrit (and every other advanced language of the world), most words carry more than one meaning. 'Ashva' here has at least three different meanings, but the ill-informed linguists that analyzed our texts over the past couple centuries only considered 'horse'.

Another meaning of 'Ashva' is the root of the herb called 'Ashvakanda' which is popularly known as 'Ashvagndha' in the modern time. In the Vedic era, many parts of this herb were used in the Havan (fire rituals) where they were offered in the fire mixed with other herbs; because their vapours are beneficial for the health of the environment and the people living in the area. Ironically, the ignorant interpreters translated the meaning of these verses as sacrifice of a horse in the sacred fire.

In the same manner 'Aja' has two meanings—one is 'goat' and the other 'elaichi' or cardamom. And, in this case too, the same confusion was created. Today, this is a well-proven fact that burning of cardamom helps prevent growth and spread of germs/pollutants in the air. The basis of these experiments was the knowledge from our elders/scriptures only. It is clear that various parts of the herbs 'Ashva' and 'Aja' were burned in the sacred fires, rather than horses and goats.

Even from the analytical perspective this interpretation is much more sensible in context of the civilization that believed that even the smallest life-form is precious and to be preserved and cherished. How could the same people be propagating animal sacrifice in their sacred fire? Especially when the effects of the two are exactly opposite on the environment. It is quite possible that these kinds of myths were knowingly circulated by the foreign invaders to put down the wisdom of our ancient system.

Coming back to the word 'Ashva', there is also a third meaning which is more fitting as part of the compound word 'Ashvamedha'. 'Ash' is a Sanskrit root which means 'to consume'; 'va' means 'to fulfil' and 'medha' means 'collectively'. When the three are combined using the rules of Sanskrit grammar, they create the word 'Ashvamedha' - which would mean 'to fulfil the collective needs of consumption'. A Yagna with this name would mean 'a mission to fulfill the needs of consumption collectively'. Most common needs of consumption of people anywhere are food, water, shelter, infrastructure etc in that order. Therefore, a Yagna that was undertaken to fulfill the basic needs of a kingdom's people, was called the 'Ashvamedha Yagna' - much like the welfare schemes of the modern-day administrations. Only difference was that the ancient kings took their mission much more seriously and often went to great lengths to complete them.

What was the Objective of Sagar's Yagna?

So what was the kind of 'Ashvamedha' that our philanthropic emperor Sagar conducted? What was the one thing that was particularly scarce for his people? These verses from 'Valmiki Ramayan' (Balkand, Chapter 31 verse 7-8) give us a hint of that -

तस्य पर्वणि तं यज्ञं यजमानस्य वासवः।। (7)
राक्षसीं तनुमास्थाय यज्ञियाश्वमपाहरत्। (8)

Meaning: the 'Ashva' of the Yagna was stolen by Indra assuming a demon form.

Since the usual form of Indra is Devta (do-gooder of society) where he brings rain - his demon form would be the exact opposite, where he inhibits it. So these verses clearly indicate that because of absence of sufficient rain, this area was suffering from scarcity of water and king Sagar decided to find ways to fulfill this scarcity. When rain from the sky is not enough, it becomes paramount to explore other sources of water. Keeping this fact in mind, Sagar (and his team of water-resources engineers), decided to explore the ground water resources first—and started looking for the 'Ashva' (the required resource) of this Yagna under the surface of Earth.

Where was This Search for Goundwater conducted?

Once again, this question is answered in these verses of 'Valmiki Ramayan' (Balkand, Chapter 31, Verses 4-5) -

शंकर–श्वशुरो नाम्ना हिमवानिति विश्रुतः।। (4)
विंध्यपर्वतमासाद्य निरीक्षेते परस्परम्।
तयोर्मध्ये समभवत् यज्ञः स पुरुषोत्तम।। (5)

Meaning: Shankar's father-in-law known by the name Himalaya, and the Vindhya Mountain, where these two see each other; in the land between them was this Yagna conducted, O Purushottam!

It is clear from these verses that even though the kingdom of Sagar extended all the way to the Indian ocean in the south, the groundwater-survey was only conducted in the area between the two mountain ranges mentioned above, which geographically translates to the great northern plane. Calculating from the map it roughly comes out to an area of 15,00,000 square kilometre. To explore such a large area obviously required a suitably large team of technicians.

Who were the 60,000 sons of Sagar?

According to the modern-day civil engineers, who have experience of exploring ground-water resources, the area to be surveyed is divided into a grid of 5km x 5km squares (25 sq km); and each square is then assigned a team of 10 workers with drilling machines. Now, if we were to calculate the total number of workers needed for 15,00,000 sq km, to effectively complete the survey of their assigned area in 5-6 months, it comes to exactly 60,000.

On the recommendation of his technical advisors, King Sagar employed exactly the same number of technicians for this mammoth task. When this Yagna (mission) was started Sagar's grandson was already in his youth, so he was definitely an old man. It would be natural for an old king to address the workers of his pet project as 'sons', which he did. And the verses with such words of address were perhaps interpreted as Sagar having suddenly acquired 60,000 sons. While it is completely illogical and impossible for a mortal man to have 60,000 sons, it is perfectly reasonable and possible for him to address 60,000 technicians as sons. Since he then instructed these 60,000 to conduct the ground-water survey, it is reasonable to assume that they were a team of experts in the said discipline, while it would be completely unreasonable to think that all 60,000 sons of a king would be inclined similarly and have knowledge of a very specific subject.

When we choose the reasonable options from the above arguments, we can safely conclude that these 60,000 were the members of the team of technicians that Sagar employed to carry out the task of finding water for his thirsty masses.

Drilling Beneath the Surface of Earth for Water

Emperor Sagar then said to them (Valmiki Ramayan, Balkand, Chapter 31, verses 14-15)

एकैकं योजनं पुत्राः विस्तारमभिगच्छत।। (14)
यावत् तुरगसंदर्शस्तावत् खनत मेदिनीम। (15)

Meaning: 'Sons, at a distance of one Yojan each, drill to the depth upto which you find water.' (Here Yojan means the side of the planned grid, say 5 kms.)

Further it is said:

योजनायामविस्तारमेकैको धरणीतलम् ।
बिभिदुः पुरुषव्याघ्रा वज्रस्पर्शसमैर्भुजैः ।। (18)

Meaning: Each one of them drilled into the Earth at a distance of one Yojan each with diamond tipped arms.

If each of the 10 members of individual teams made 10 drill-holes, then they each got one Yojan land in their share. Vajra-Sparsh directly translates to 'Diamond tip of the drilling arm'. It is very clear from this one verse that the engineers used these kinds of drilling machines to drill to a depth upto which water was available. (the presence of aquefers found)

In this manner when the 60,000 technicians of king Sagar surveyed the northern plains in search of ground-water they did not find it enough for fulfilling the needs of his masses - in other words the 'Ashva' of the Yajna was not yet found.

Exploring the North-Eastern Mountains for Water

Another option for finding the required water was surface water, for which the same set of 60,000 technicians were ordained. They scanned all directions and finally located some in the North-east, the land of the Brahmaputra. (According to the Pauranic tale, they found the horse tied in the hermitage of sage Kapil). Once again, if we analyze the symbolism here to find the logical meaning, while 'Kapil' was an ancient sage, the word 'Kapil' also means brown. When we substitute the second meaning instead of the first the event now makes more geographical sense—the team of engineers located 'enough water' (Ashva) in the land of 'brown mountains' of Brahmaputra catchment (Kapil's hermitage); and for the purpose of harvesting it they went to this area equipped with a different set of tools - excavators, rock-excavators, tree-saplings and rocks. (Valmiki Ramayan, Balkand, Chapter 40, Verse 27)

खनित्र लांगलधरा नानावृक्ष शिलाधराः ।

Our interpretation is further supported by the choice of tools here. While they had used diamond-tipped drilling machines to reach water underground, they were now using 1) rock-excavators to cut through the mountain, 2) trees for soil conservation in the catchment area and 3) rocks for making a series of big and small dams. It is quite clear that they were not planning to capture a horse.

But they once again failed in their mission here. As is said in the Pauranic tale - they were impatient and rude and abused the meditating sage Kapil for having their horse and the irritated sage waking up from his trance of many years and in his ire causing the death of the thousands of sons of Sagar. A logical interpretation of this symbolic sentence would be that the technicians were not careful as they started to dig through the brown mountain region, which had long been undisturbed by humans. In their hurried attempts they caused a destabilization of the region and many of them died as a result of it; most probably a land-slide.

In this manner, Sagar had to relinquish this quest too. When so many of the scientists had died without a successful completion of their assigned mission, it is obvious that their souls would only find peace with its completion. Not just in this story, that is how people have paid tribute to their predecessors in every part of the world. Therefore, Sagar continued to make attempts to locate other sources of water but failed every time. He could not see his efforts bear fruit during his lifetime.

Attempts Made by the Descendents of Sagar

After the demise of Sagar, his Yagna (mission) was taken over by his descendents. Anshuman, the grandson of Sagar, started a quest and found the 'Ashva' (required amount of water) in the area of Mansarovar Lake, but he could not find a way to bring it down to his masses. The pauranic tale tells us that he spent 32 years in the Himalayas for this purpose doing extreme 'Tapasya'. Once again, the word 'Tapasya' does not mean any kind of spiritual austerities observed by standing on one leg and chanting Mantras for years; as such a practice would never yield results. Instead, it means hard work; which is known to always yield results. And this word has been used in that very context in this tale; not once or twice but multiple times. When the 'Tapasya' (hard work) of Anshuman didn't succeed, his son Dileep took over his quest and he too failed despite trying to accomplish it through various means, as he could not come up with a viable plan. Eventually, his son Bhagirath was able to find a solution to this long-standing woe of their people.

How did Bhagirath Complete Sagar's Yagna?

The Pauranic tale further says that after doing 'Tapasya' for many years Bhagirath was granted a boon by Lord Brahma that He will ask the celestial River Ganga to leave her abode in heaven and flow to Earth instead for the noble task of washing the sins of masses. The only scientific interpretation can be that after many years of hardwork in the Himalayan region, Bhagirath finally found an engineer (as creator Brahma is the creating power of universe) who promised to bring down the waters from the sacred region of Mansarovar down to the northern plain.

To collect the water in this region, they first planted trees in an extended region of Himalayas, which acted as hair-locks of Lord-Shiva.

In this manner, Bhagirath with the help of many others created a reservoir called 'Bindusar' and brought down River Ganga down to the thirsty planes. The Pauranic tale says that after being disciplined by Lord Mahadev, Ganga told Bhagirath to lead the way and that she walked behind him, flowing over regions that he wanted her to. Even the most stubborn among our readers would agree that Ganga was never a woman nor was she a goddess, it is just a name given to a river. And for a river to follow the orders of a man, any man, would only mean that he was able to control the path and direction of the water-flow through various engineering techniques.

The tale further says that through the touch of the celestial waters, the spirits of the 60,000 dead sons of Sagar, finally found peace. We all know that Ganga flows all the way to the eastern end of the plains, where she finally meets with Brahmputra to create the largest delta on Earth. It is the same region where the team of technicians had perished while trying to harvest its water. And that winds up the story in a beautiful manner, as the souls of ancestors finally found peace upon the completion of their collective dream.

Conclusion

And this is how the region between the great Himalayas to the North and Vindhyachal to the south, was supplied enough water in the form of River Ganga. It was by no means an easy task, from Sagar to Asmanjas to Anshuman to Dileep to Bhagirath, it took five generations of kings committed to this quest. The promise of his great great grandfather was fulfilled by Bhagirath after four generations of failed attempts. All these philanthropic kings experimented using their own brainpower and techniques and, in the end, one was successful.

As was said in the beginning, Sagar ascended the throne in 6797 BCE, the descent of Ganga should be dated to approximately 6600 BCE. From then to now it has been 8600 years and the celestial waters still flow uninterrupted through our northern plains, bringing a new lease of life to the people of that region.

In a similar manner Saraswati too was brought down from the peaks of Himalaya. It is a known historical fact that it was mainly on the banks of these two rivers (Saraswati and Ganga) that the major centres of our ancient civilization developed and flourished for many millennia before we were invaded and enslaved by foreigners a thousand years ago. It is definitely a point to ponder that if our ancestors were advanced enough to design and direct the paths of these great rivers, how advanced must be their knowledge of science and how great their building skills. And above even those - their compelling attitude of philanthropy, that made them give up the comforts of their palaces and drove them into the harsh mountains so that their people could live in better comfort. Not just one but five generations of these kings worked tirelessly for the singular selfless purpose of serving the masses.

CHAPTER 18

Astronomy

It is a well-known fact that ancient Indian astronomers had very advanced knowledge of Astronomy. They could make accurate calculations of the movement of planets and stars, and movement of the orbits of the planets and their apices.

As the movement of planets is elliptical, there are two apices of the elliptical orbits. In Indian Astronomy, these apices are called उच्च and मन्द (शीघ्र) of the planets.

The movement of the apices of the orbits is so slow, that it may take millions of years to complete one revolution. In fact, the apex of Saturn's orbit takes about 110 million years to complete one revolution or about 300 thousand years to move by one degree. The calculation of these movements involves solving of very intricate differential equations, which would not have been possible without the knowledge of differential and integral calculus.

The observation of such slow motion could not be possible without the use of highly sophisticated telescopes.

As per the Indian Astronomy, the formation of the planets of the solar system started taking place about 1.973 billion years ago and it took about 17 million years in the formation of the planets. Thus, the age of the Earth was estimated to be 1.956 billion years.

The estimate of the age of the Earth is not very widely off the mark when compared with the estimate of the modern scientists. However, the modern scientists have not so far ventured into estimating the time taken in the formation of the planets.

The smallest unit of time used by the Indian Astronomers was a 'Truti', which was equal to about 0.308 microseconds. The largest unit was a 'Kalpa' which is equal to 4.32 billion years. Brahma's age was estimated to be 72,000 'Kalpas' or about 3,110 trillion years.

CHAPTER 19

Cosmology

According to modern science, the Universe has evolved from the space-time continuum. As per Einstein's theory, curvature in the space-time continuum appears as matter. Thus, matter has no existence independent of the space-time continuum. The term equivalent to space-time continuum in Sanskrit is 'देश–काल'.

However, in the ancient Sanskrit literature, this term is rarely used as an independent entity, and instead, the term 'देश–कालात्म' is frequently used. 'देश–कालात्म' means the continuum of space, time and Ātman. This term has been used in *Charaka Samhita*, the famous treatise on Ayurveda, wherein Charaka says that the knowledge of the fundamental science of 'देश–कालात्म' is more effective in combating mental diseases than the use of medicines. This term has also been used in *Bhāgavata Purāna* to indicate that a knowledgeable person is supposed to have the knowledge of the fundamental science involving space-time-Ātman.

Thus, the modern scientists as well as the ancient Indian scientists are on the same page on the issue of the Universe being evolved from the space-time continuum. However, the major difference is that the ancient Indian scientists go one step further when they say that the space-time continuum is not the ultimate substance from which the Universe has evolved. Rather, the ultimate basis is the Ātman from which space-time has evolved.

The Sanskrit equivalent of the word 'curvature' is 'ग्रंथि', which, in layman's terms, means a 'knot'. Our sage scientists say that the Universe is actually the result of a 'ग्रंथि' (curvature) in the Mind, just like modern science says matter is nothing but a curvature in the space-time continuum. There are four intermediate steps in the process of evolution of the space-time from Ātman, viz. Ego (super ego), Intellect, Mind, and Sense Organs.

The basic concept of cosmology is elucidated in the 7th chapter of the Gita, wherein Lord Krishna says to Arjun,

ज्ञानं तेऽहं सविज्ञानमिदं वक्ष्याम्यशेषतः।

Meaning:"I shall tell you this complete knowledge along with Science."

And then says,

भूमिरापोऽनलो वायुः खं मनो बुद्धिरेव च।
अहंकार इतीयं में भिन्ना प्रकृतिरष्टधा।।
अपरेयमितस्त्वन्यां प्रकृतिं विद्धि मे पराम्।
जीवभूतां महाबाहो ययेदं धार्यते जगत्।।
एतद्योनीनि भूतानि सर्वाणीत्युपधारय।
अहं कृत्स्नस्य जगतः प्रभवः प्रलयस्तथा।।
मत्तः परतरं नान्यत्किंचिदस्ति धनंजय।
मयि सर्वमिदं प्रोतं सूत्रे मणिगणा इव।।

Meaning: "Matter, the seed particles of matter, Energy, Waves, Space, Mind, Intellect and Ego – this is the varied eightfold Nature of Mine (the Supreme Ātman). Know that there is another Nature of Mine, subtler than this, which is in the form of Jeeva (life-force), which sustains this Universe.

Understand that it is somewhat like this that all beings are the Creation of this (subtler Nature).

I, (the Supreme Ātman) am the Originator as well as the Dissolutor of this whole Universe. Arjuna, there is nothing else beyond Me (the Supreme Ātman). All this is stringed in Me just like jewels in a string."

These shlokas read along with the following shloka of the 3rd chapter of the Gita -

इन्द्रियाणि परायण्याहुरिन्द्रियेभ्यः परं मनः।
मनसस्तु परा बद्धिर्यो बुद्धेः परतस्तु सः।।

Meaning: "The organs are said to be subtle mind is subtler than organs, intellect is subtler than mind, and that which is beyond intellect is That (Self)." This gives the complete gist of ancient Indian cosmology.

It may be observed that usually the words used for the five elements here, viz. भूमि, आपः, अनल, वायु, and आकाश are mistranslated. On going through the Sanskrit books dealing with Science and Technology, it becomes clear that 'वायु' means 'wave'. Similarly, अनल stands for energy. In Vedic literature, the words 'तेजस्' and 'वर्चस्' are often used as synonyms of 'अग्निः'. 'वर्चस्' means energy, and thus the element 'अनल' is Energy. Similarly, 'आपः' means the seed particles of matter, and 'भूमि'

means matter. The eight elements from भूमि to अहंकार have been given in a sequence of their evolution in reverse order.

The element 'Apah' is described in the ancient Sanskrit book "Anshu-bodhini" (अंशु–बोधिनी) under सूत्र 22, it is stated that

अब्रूपं द्विविधं प्रोक्तं विद्युद्भूतप्रभेदतः।
वैद्युतोदकमुद्भूतमव्यक्तावरणोपरि।।
यदेतस्मादभूत्सृष्ट्यां सृष्टिबीज विभागतः।
तद्भूतोदकमित्याहुर्ज्ञानिनस्तत्त्वदर्शिनः।।

Meaning 'Ap' is of two types - electricity and the element.

The 'electricity' form of 'Ap' originated on the cover (surface) of the unmanifest. That which originated from this, as the seed (particle) of matter, is called the element 'Ap'.

Thus the element 'Apah' (plural of 'Ap') means the seed particles of matter, which maybe electrons and positrons or some other basic particles.

It may be observed that the book 'Anshubodhini' deals exclusively with the characteristics and applications of electromagnetic waves. This book contains 12 chapters and 1000 aphorisms (सूत्र). Unfortunately, only the first chapter of this book containing 50 सूत्र is currently available in India.

Under सूत्र 3 of this book, the range of electromagnetic microwaves, generating intense heat is given as 4.7 GHz to 3.7 MHz, which is very close to 5 GHz to 5 MHz range given by the modern science. The relevant shlokas are given below -

शतकोट्यष्ट लक्ष त्रिसहस्रत्रिशतं क्रमात् ।
निमेष मात्राद् भ्रमण मग्निशक्तेस्स्वभावतः ।।
तथैव सोमशक्तेश्च प्रभवत्यतिवेगतः ।

Meaning: "The frequency of अग्नि–शक्ति waves is in the range of hundred crore cycles per निमेष, to eight lac cycles per निमेष, and that of सोम–शक्ति waves is in the range of three thousand per निमेष to three hundred cycles per निमेष."

It has been established in chapter 1 that one निमेष = 0.2127508 sec

Therefore the frequency of अग्नि–शक्ति waves would work out to be 4.7 GHz to 3.7 MHz. And that of सोम–शक्ति waves works out as 14,000 to 1,400 Hz.

In the moden science no equivalent of सोम–शक्ति waves is available. However in another ancient Sanskrit book 'बृहद् विमान शास्त्र', it is indicated that the सोम–शक्ति waves were used to create a cover around the aeroplane to make it invisible.

In 'अंशु–बोधिनी' it is stated that a combination (intermingling) of अग्नि–शक्ति waves and सोम–शक्ति waves can generate consciousness in the bodies.

Research on the application of these two categories of electromagnetic waves can lead to new vistas of discovery in modern science.

Thus, ancient Indian cosmology is a synthesis of fundamental Physics, Psychology, Parapsychology, and a step beyond Parapsychology.

It has been proved in a famous experiment in modern science that "an electron occasionally moves beyond an electrical barrier, for a very short period, but, if observed in its new position, it does not move back to its original position." Thus, the existential position of a particle is dependent on its observer. A logical inference of this experiment would be that the existence of the observable Universe is not independent of the observer (साक्षी). The day is not far when the modern scientists will reach a conclusion which is very akin to the theory of cosmology evolved by the ancient Indian sage scientists.

PART-IV

History and Prehistoric History

CHAPTER 20

Dating of Ancient Indian History

The Greek Historian Strabo has recorded that Megasthenes was sent to Pātaliputra as an ambassador to 'Sandrakottos', and Deimachus as an ambassador to his son, 'Amitrachades' [ref: 1, p18,19]. In 1793 CE, Sir William Jones, the founder of the Royal Asiatic Society presented to the Asiatic Society his "discovery" that 'Sandrakottos' was the Indian king "Chandragupta Maurya". This "discovery" by him was readily accepted by the modern historians, although there is very little resemblance between the names 'Sandrakottos' and 'Chandragupta', and 'Kottos' especially has little resemblance to 'Gupta'. Moreover, the name of Chandragupta Maurya's son could not even be remotely related with 'Amitrachades' as recorded by the Greek historians. It may be mentioned that to "overcome" this discrepancy, the modern historians have introduced in the history books, that Chandragupta Maurya's son, Bindusāra, used the title "Amitraghāta". There is no inscriptional or literary evidence to this effect [Ref: 2]. Anyway, "Amitraghāta" means "Slaughter of enemies", not "Slayer of enemies", and thus, it cannot be the title of a person as claimed by modern historians. Also, the period of Alexander and Seleucus does not even remotely match the period of Chandragupta Maurya, as per the chronology of kings given in the *Bhāgavata Purāna*. Then, who was Sandrakottos?

In Greek mythology, 'Kottos' was a giant with 50 heads (100 ears) and 100 hands. Several kings of the Andhra dynasty which ruled over India, with their Capital at Pataliputra, for 456 years, [Ref. *Bhāgavata Purāna* Skandh 12, Chapter 1] before the Gupta dynasty used the title 'Shatakarni'. Shatakarni literally means "a person having a hundred ears".

The Greek Historians may have identified the surname 'Shatakarni' with the Greek 'Kottos', who also had a hundred ears. Thus, 'Kottos' is a Greek equivalent for the Indian 'Shatakarni'. One of these Andhra dynasty kings who wrote 'Shatakarni' with his name was 'Sundara' [Ref: Brahmanda Purāna & Vishnu Purāna]. 'Sundara' very much resembles 'Sandra'. It indicates that 'Sandrakottos' of the Greek historians was king 'Sundara Shatakarni' of the Andhra dynasty. In *Bhāgavata Purāna,* his name has been mentioned as Sunandana. It may be noted that in Sanskrit, 'Sunandana' is a synonym of 'Sundara'. Another synonym of 'Sundara' is Priyadarshi. In the classical Hindi and Sanskrit literature, and perhaps in Greek literature too, it was a common practice to substitute the names of persons and cities with their synonyms. In the latter part of the Shloka where 'Sunandana' is mentioned, it is written in the *Bhāgavata Purāna,*

चकोरा बहवो यत्र शिवस्वातिररिन्दमः [Bhāgavata Purāna, Skandha 12, Chapter 1]

which literally means, Shivasvāti Arindama was where Greek partridges (Chukar partridges) were abundant. It may be noted that in *Bhāgavata Purāna,* the word 'चकोरा' has been misprinted as 'चकोरो', which is not correct, because with the adjective 'बहवो यत्र', which is in plural, the noun should also be plural, whereas 'चकोरो' is singular. So, it must be read as 'चकोरा', otherwise the whole clause would become meaningless. Due to this misprint, 'चकोर' has been misinterpreted by some as the name of a king.

As Chukar partridges are abundant in the region ranging from North East Greece to North West India, [Ref: 3] the statement indicates that Sunandana's (Sundara's) brave son, Shivasvāti, nicknamed 'Arindama' meaning "Thrasher of the enemies", was deputed as a 'Satrap' in the North West region of the Indian kingdom, perhaps because that region of the kingdom was prone to invasion by the Greeks. It may be noted that since 'Arindama' is a synonym of 'Amitrachhid' in Sanskrit, both meaning 'Thrasher of the enemies', he was also nicknamed as 'Amitrachhid', which may have been recorded as 'Amitrachades' by the Greek historians. Thus, the identification of the father-son due viz. 'Sandrakottos' and 'Amitrachades', with 'Sundara Shatakarni' and 'Arindama' is complete.

To establish this identification further, let us look into the time-periods of Megasthenes and Sundara Shatakarni. As per Greek history,

Alexander invaded India in 326 BCE. After his death in 323 BCE, his empire was divided into several parts, ruled over by different kings. The Eastern part close to the Indian kingdom was ruled by Seleucus. Subsequently, Seleucus entered into a treaty with the Indian king in 302 BCE in which he gave his daughter, in marriage, to the Indian king along with a large part of his kingdom, and in return, got 500 elephants from the Indian king who had 9,000 elephants.

Seleucus sent Megasthenes as an ambassador to the Indian king at Pataliputra. The Western historian C. Miller writes that, "The time when he discharged his embassy or embassies and how long he stayed in India cannot be determined, but he was probably sent after the treaty had been struck and friendship had sprung up between the two kings. If, therefore, we make the reign of Sandrakottos extend to the year 288 BCE, Megasthenes would have set out for Pataliputra between 302 BCE and 288 BCE." [Ref: 1, p14]

As per the chronology of dynasties of Indian kings after the Mahābhārata War given in *Bhāgavata Purāna*, the Andhra dynasty ruled over India for 456 years, from 610 BCE to 154 BCE (for detailed calculations, see sections 2.0 and 3.0

below). The separate period of reign of each king has not been given in *Bhāgavata Purāna,* but in *Matysa Purāna,* the separate periods of reign of these kings have been mentioned. From this, the year of the start of Sundara's reign works out as 292 BCE (see section 3.0), which tallies very well with the period mentioned by C. Miller, the Western historian.

The above analysis indicates that 'Sandrakottos' of the Greek historians could be none other than Sundara Shatakarni of the Andhra dynasty, as all the facts, viz. his name, his son's name, and the period of his reign tally exactly with those given by Greek historians and Miller. It may be noted that Max Muller pointed out, on the basis of the records of the ancient Greek writers that "Sandrakottos" established his new kingdom only after the invasion by Alexander [Ref: 4]. That means 'Sandrakottos' was not the Indian king when Alexander invaded India in 326 BCE.

Some Indian researchers have identified 'Sandrakottos' with the Gupta king 'Chandragupta I' [Ref: 2] or Samudragupta [Ref: 7] which does not appear likely, because the reign of the Gupta dynasty started much later than the period of

Megasthenes. If the period of Chandragupta I or Samudragupta is pre-dated to match with that of Megasthenes, then who started the Vikrami Samvat in 58 BCE? Thus, the identification of 'Sandrakottos' with Chandragupta I or Samudragupta, though a well-intentioned attempt, cannot be accepted as correct due to this chronological anomaly. Further, their names do not have much resemblance with 'Sandrakottos', and 'Kottos' in particular has little resemblance to 'Gupta'.

2.0 Year of the Mahābhārata War

The following three independent sources establish the year of the Mahābhārata War -

1) The Aihole Inscription: The Famous Aihole Inscription, established in the 7th century CE by king Pulikeshi II of the Chalukya dynasty reads :

त्रिंशत्सु त्रिसहस्रेषु भारतादाहवादितः।
सप्ताब्दशतयुक्तेषु गतेष्वब्दषु पंचसु।।
पंचाशत्सु कलौ काले षट्सु पंच शतेषु च।
समासु समतीतासु शकानामपि भूभुजाम्।।

Meaning: "To date, 3,735 years have elapsed since the Mahābhārata War, and 556 years have elapsed since the start of Shaka kings' rule in Kaliyuga."

From this, it is evident that the Mahābhārata War took place (3735 - 556) = 3179 years before the 'Shaka' Samvat which started in 78 CE, giving the year of Mahābhārata as 3102 BCE.

2) Internal Evidence in the Mahābhārata Itself:

A shloka in the Mahābhārata reads:

प्रयाते तु ततः कर्णे योधेषु मुदितेषु च।
निःसरंतो व्यदृश्यन्त सूर्यात् सप्त महा ग्रहाः।।

[Parva 8, Chapter 37]

Meaning: "When Karna, along with joyful warriors, marched (for the war), seven major planets appeared to be coming out from the Sun."

This shloka refers to the time when Karna was the Commander-in-Chief of the Kaurava army on the 17th day of the war. It is clear from this verse that seven planets were in near alignment with the Sun at that time.

According to the famous book on Astronomy "Sūrya Siddhanta", this rare event occurs at the start of Treta yuga or at the start of Kaliyuga. It indicates that Kaliyuga started on or near about the 17th day of the Mahābhārata War, thus establishing that the Mahābhārata War took place at the start of Kaliyuga, i.e., in 3102 BCE, as Kaliyuga started in 3102 BCE [Ref: 8, 9]

3) Ain-e-Akbari (आईने अकबरी): In "Ain-e-Akbari", the famous treatise on Indian History, written by the Prime Minister of king Akbar, it is recorded that "At the very start of Kaliyuga, the first king was Yudhishthira. Before the start of Vikrami Samvat, 3044 years of Yudhisthira era had elapsed." [Ain-e Akbari, published in 1867 CE, p269]. Thus, Yudhishthira's Coronation, at the end of the Mahābhārata War took place in (58+3044) = 3102 BCE, and also that Kaliyuga started in 3102 BCE.

From this, it is established on the basis of three independent sources, that the Mahābhārata War took place in 3102 BCE, and that Kaliyuga also started in 3102 BCE.

3.0 Dating of Post-Mahābhārata History

In *Bhāgavata Purāna* [Skandha 12, Chapter 2, Shloka 1], it is written that 1,510 years had elapsed from the birth of Parikshit upto the coronation of Nanda. As Parikshit was born a few months after the Mahābhārata War, in 3102 BCE, the year of coronation of Nanda works out to be (3102-1510) = 1592 BCE.

The period 1,510 has been wrongly interpreted by some as 1,115 years because the printed wording is "एतद् वर्ष सहस्रं तु शतं पंच दशोत्तरं" instead of "एतद् वर्ष सहस्रं तुशत पंच दशोत्तरं". In this context, it may be observed that in *Bhāgavata Purāna* itself, it is mentioned that after Sahdeva, the son of Jarāsandha, who was installed as the king a few years before the Mahābhārata War, his descendents ruled for a thousand years [Ref: Skandha 9, Chapter 22, shlokas 46 to 49], and after that, five Pradyotans ruled for 138 years and 10 Shishunagas ruled for 360 years [Ref: Skandha 12, Chapter 1, shlokas 2 to 8], adding up to (1000+138+360) = 1,498 years. The period of a "thousand years" may be a rounded off figure for 1,012 years. Thus, the correct version is 1,510 years and not 1,115 years.

After Nanda's coronation, he and his eight sons ruled for a hundred years [Ref: Bhāgavata Purāna, Skandha 12, Chapter 1, shlokas 9 to 11], and then Chandragupta Maurya was installed as the king [Ref: Bhāgavata Purāna, Skandha 12,

Chapter 1, shlokas 12, 13]. Thus, the year of coronation of Chandragupta Maurya works out to be (1592-100) = 1492 BCE.

The names of contemporary kings found on Ashokan inscriptions are Amtiyoka, Tulamaya, etc. Amtiyoka ruled Afghanistan at around 1475 BCE, which then appears to be the approximate date of Ashoka's - the grandson of Chandragupta Maurya's - rule [Ref: 10].

According to Vedveer Arya, Chandragupta Maurya ascended the throne in 1534 BCE [Ref: 7].

According to Swami Prakashananda Sarasvati, Chandragupta Maurya ascended the throne in 1541 BCE [Ref: 11].

The main discrepancy in their calculations has crept in due to misinterpretation of the following shlokas in *Bhāgavata Purāna* (skandha 12, chapter 3):

विष्णोर्भगवतो भानुः कृष्णाख्योऽसौ दिवं गतः।
तदाविशत् कलिर्लोकं पापे यद् रमते जनः।। २९।।
यावत् स पादपद्माभ्यां स्पृशन्नास्ते रमापतिः।
तावत् कलिर्वै पृथिवीं पराक्रान्तुं न चाशकत्।। ३०।।
यस्मिन् कृष्णो दिवं यातः तस्मिन् एव तदाहनि।
प्रतिपन्नं कलियुगं इति प्राहुः पुराविदः।। ३३।।

Meaning: "When That Lord Vishnu's Light named Krishna left for Heaven, then Kali entered the world so that people's minds take pleasure in sinful activities. So long as Lord Krishna kept touching the Earth with his lotus feet, till then, Kali could not exert its power on the Earth. The day Lord Krishna left for heaven, on that very day, Kaliyuga started to act - so say the Historians." The connotation of these shlokas is very clear. It means that Kaliyuga (although it had started earlier) could not exert its power on the minds of the people until Lord Krishna left for heaven. Some people have interpreted these shlokas to mean that Kaliyuga started on the day Lord Krishna left for heaven. This misinterpretation pushes back the year of Mahābharata and other events of history by 36 years.

The dating of Chandragupta Maurya by the Western Historians as about 320 BCE is off by about 1,170 years. As the Western Historians have taken the year of Chandragupta Maurya's coronation as the sheet anchor of Indian History, the whole History of India has been grossly distorted.

After that, the periods of reign of the following dynasties have been recorded in the *Bhāgavata*

Purāna, Skandha 12, Chapter 1, shlokas 13 to 28.

1. Maurya 337 years
2. Shunga 200 years
3. Kanvayan 345 years
4. Andhra 456 years

Thus, the year of the start of the Andhra dynasty rule, works out to be (1492-337-200-345) = 610 BCE.

It may be observed here that in the *Bhāgavata Purāna*, the years of reign of the Maurya dynasty has been misprinted as 137 years instead of 337 years. The number of kings of the Maurya dynasty has been mentioned as 10. In the *Matsya Purāna*, the total years of reign of 6 Maurya kings is given as 199 years. In view of that, 137 years is clearly a misprint. The misprinted word is "सप्तत्रिंशच्छतोत्तरं" meaning 137, instead of "सप्तत्रिंशच्छतत्रयं" meaning 337. It may be noted that both these words have the same number of syllables and sound alike.

Similarly, the number of years of reign of the Shunga dynasty kings has been misprinted as "वर्षशताधिकं" which means "more than a hundred years". (How many more than a hundred?) Thus, this word does not convey any specific sense. The nearest sounding word with the same number of syllables is "वर्षशतद्वयं", meaning 200 years.

The total number of years of reign of the Andhra dynasty kings, as per *Matsya Purāna,* upto the start of Sundara's reign, adds up to 318 years. Thus, the year of Sundara's coronation works out to be (610-318) = 292 BCE, which tallies very well with the period (given by Western Historian, C. Miller) when Megasthenes was sent as an ambassador to the Indian king sometime between 302 BCE and 288 BCE.

As per the *Bhāgavata Purāṇa,* after the Andhra dynasty, Abhiras (Gupta Dynasty) took over. The year of the end of the Andhra dynasty and the start of the Gupta dynasty works out to be (610-456) = 154 BCE. As per *Bhavishya Purāna,* Vikramaditya ascended the throne in 85 BCE. He started Vikrami Samvat in the year 58 BCE. Thus, the period of reign of the first two kings of the Gupta dynasty works out to be 69 years.

4.0 Dating of the Pre-Mahābhārata History

Megasthenes has recorded in his *Indika* [Solin 52.5] that "Father Bacchus was the first to invade India, and he was the first of all those who triumphed over the vanquished Indians. From him to Alexander the Great, 6,451 years are reckoned with three months additional, the calculation being made by counting the kings who reigned in the intermediate period, to the number of 153." [Ref: 1, p116].

The 153 kings mentioned here are Indian kings, as is clear from his other statement in Frag. L.C. [Plin Hist. Nat. VI, xxi, 4.5], "For the Indians stand almost alone among the nations in never having migrated from their own country. From the days of Father Bacchus to Alexander the Great, their kings are reckoned at 154, whose reigns extend over 6,451 years and 3 months." [Ref: 1, p115-116]

Here, the 154th Indian king was the one who was still ruling when Alexander was installed as the king of Greece, and therefore, duration of his reign has not been included while counting the years of reign of the 153 kings who reigned in the intermediate period.

The above-cited statement of Megasthenes may be read along with the statement in *Harivansha Purāna* [Part 2, Chapter 13, shlokas 30, 31], "Ruruka's son was Vrika, and 'Bahu' was born to Vrika. The Haihayas and Talajanghas, in alliance with the Scythians, Greeks, Combodians, Pāradas and Persians dethroned him."

This narration perfectly matches with the statement recorded by the Greek historian Megasthenes. Reading the two records together, it is evident that perhaps the Greek king Bacchus (in alliance with others), had invaded and defeated the Indian king Bāhu, father of Sagara who was the famous ancestor of Lord Rāma.

From the statement of Megasthenes, we can find the year of coronation of king Sagara as follows. Alexander was installed as the king in 332 BCE. At that time, as per the number of years of reign of the kings of the Andhra dynasty, as given in *Matsya Purān*, Riktavarna, of the Andhra dynasty, was the king of India, and was installed as the king in 346 BCE. He ruled from 346 BCE to 321 BCE. Thus, as per Megasthenes' record, 6,451 years (ignoring the additional three months) had elapsed from the coronation of king Sagara upto the coronation of Riktavarna. Thus, the year of coronation of king Sagara works out to (346+6451) = 6797 BCE.

It may be noted that, as per the chronology of kings as given in the Purānas, the number of kings from Sagara upto Riktavarna, work out to be 153, which is the same as mentioned by Megasthenes. The details are as under:

1. Number of kings of Sūrya dynasty from Sagara, upto the Mahābhārata [Ref: Harivansha Purāna] = 52
2. Number of kings of Brahadratha dynasty, after Mahābhārata [Ref: Matsya Purāna, chapter 271, shloka 30] = 32
3. Number of kings of Pradyota dynasty [Ref: Bhāgavata Purāna, Skandha 12, chapter 1] = 5
4. Number of kings of Shishūnāga dynasty [Ref: Bhāgavata Purāna, Skandha 12, chapter 1] = 10
5. Number of kings of Nanda dynasty [Ref: Bhāgavata Purāna, Skandha 12, chapter 1] = 9
6. Number of kings of Maurya dynasty [Ref: Bhāgavata Purāna, Skandha 12, chapter 1] = 10
7. Number of kings of Shunga dynasty [Ref: Bhāgavata Purāna, Skandha 12, chapter 1] = 10

8. Number of kings of Kanva dynasty = 10
9. Number of kings of Andhra dynasty upto Riktavarna [Ref: Matsya Purāna] = 15

Total = (52+32+5+10+9+10+10+10+15) = 153

The number of kings of the Kanva dynasty has not been mentioned, but the years of reign of this dynasty has been given as 345 in *Bhāgavata Purāna,* and only four names have been given.

It may be noted that, in the *Bhāgavata Purāna,* the total number of kings in the Andhra dynasty is given as 30, but only 22 names have been given. Thus, eight names are missing. Similarly, in Kanva dynasty, some names must be missing, as the average period of reign of each king in this dynasty cannot be believed to be 86 years. Therefore, the number of kings in the dynasty has been assumed to be 10 (giving the average number of years of reign as 34.5 years).

From this, the dating of Pre-Mahābhārata history can be estimated as below:

As per the chronology of the kings of Sūrya dynasty given in *Harivansha Purāna,* 52 kings had ruled from Sagara upto the Mahābhārata.

Thus, the average period of reign of each king works out to be (6797-3102)/52 = 4695/52 ≈ 71 years.

This average is not excessive in view of the fact that in the Pre-Mahābhārata era, it was quite usual that the kings ruled up to their ripe old age, and would hand over the reigns to their grandsons or even great-grandsons. Yudhisthira was over a hundred years old when he handed over the reigns to Parikshit, the grandson of Arjuna. Also, after Vasudeva, his great-great-grandson, Vajra, was installed as the king.

The number of kings from Lord Rāma upto Mahābhārata = 28. Thus, the approximate year of coronation of Lord Rāma works out to be = [3102+(28x71)] = (3102+1988) = 5090 BCE. This tallies very well with the dating of Rāma as scientifically established by Pushkar from the position of the planets and stars as given in Vālmīki Rāmayana. [Ref: 5].

The year of birth of Lord Rāma has been established by Pushkar as 5114 BCE.

Number of kings from Vaivasvata Manu to Mahābhārata = 87.

Therefore, the approximate year of coronation of Manu = [3102+(87x71)] = (3102+6177) = 9289 BCE, say 9300 BCE.

Note that the previous ice age ended in about 11,500 BP (or 9500 BCE) as per modern scientific evidence [Ref: 6, p 65].

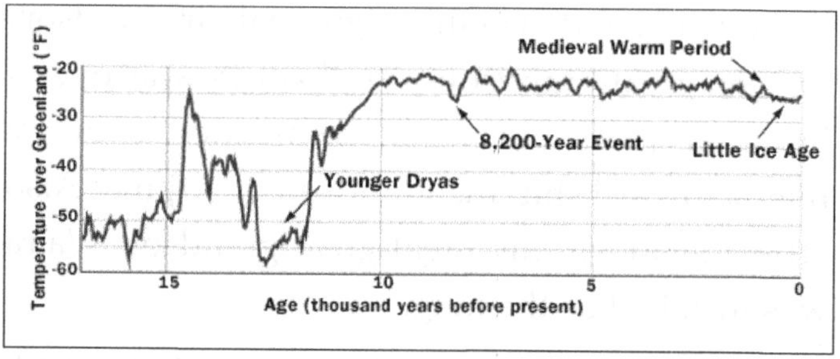

Similarly, other dates have been calculated and are given below :

Building of Ayodhya ≈ 9300 BCE

The first (आदि) Vasistha ≈ 9,300 BCE

Building of Kushasthali (which was later renovated and expanded as Dvārika) ≈ 9150 BCE

King Raivata/Revati/Space Travel ≈ 9050 BCE

Building of Kāshi/Vārānasi ≈ 8,800 BCE

Establishment of Gandhāra State (by Druhyu's descendent) ≈ 8,800 BCE

Shaunaka, the Vedic Sage ≈ 8,700 BCE

Dhanvantri, the propagator of Ayurveda ≈ 8,600 BCE

Building of Māhishmati ≈ 8,500 BCE

Establishment of Kerala, 'Kola', 'Chola' and Pāndya States (by Turvasu's descendents) ≈ 8,300 BCE

The first (आदि) Vishvāmitra ≈ 8,300 BCE

Kārtavirya Arjuna, the owner of 400m long solar Aeroplane ≈ 8200 BCE

Kanva/Medhatithi, the Vedic Sages ≈ 7,900 BCE

Shakuntala/Bharata ≈ 7,800 BCE

Trishanku ≈ 7350 BCE

Harishchandra ≈ 7300 BCE

Bhagiratha/Descendence of Bhagirathi, tributary of Ganga ≈ 6600 BCE

(The year of descendence of Bhagirathi can be scientifically estimated by taking samples from below its bed.)

CHAPTER 21

Pre-Historic History

In the last chapter, the dating of ancient Indian history has been outlined upto Vaivasvata Manu. However, Vaivasvata Manu was not the first king in ancient India. It is said that before Vaivasvata Manu, there were at least six more Manus, namely Svāyambhuva Manu, Svārochisha Manu, Uttama Manu, Tāmasa Manu, Raivata Manu and Chākshusha Manu. However, only scanty history of ancient India before Vaivasvata Manu is available in the History books, although the heirarchy of some kings in the dynasty of Chākshusha Manu is given in *Harivansha Purāna* [Book I, Chapters 2, 3]. Therein, the names of nine descendants of Chākshusha Manu, upto Daskha, have been mentioned. King Kashyap was a son-in-law of Daksha. It has also been mentioned that the 4th descendant of Chākshusha Manu, king Prithu, was the first king who ruled over the whole Earth.

Daksha had several daughters, some of whom were married to king Kashyapa, prominent among them being Danu, Diti, Aditi and Surabhi. Vasu, another daughter of Daksha, was married to Dharma. One of their sons was Prabhāsa, whose descendant was the famous engineer Vishvakarmā who developed many technologies and designs of aeroplanes.

The descendants of Danu were called Dānavas, prominent among them being Maya (मय), another famous engineer, who, besides developing several technologies and designing aeroplanes, was also a prominent astronomer. The descendants of Diti were called Daityas, prominent among them being Hiranyakashipu, Prahlāda, and Bali. The descendants of Surabhi were the Rudras. The descendants of Aditi included the Sūrya dynasty.

A great War, known as 'Star War' (तारकामय युद्ध) started, with the descendants of Danu and Diti on one side, and the descendants of Aditi, Surabhi and Vasu on the other side. This great war, included 'Space War' as the name indicates.

Besides the seven Manus mentioned above, there is the mention of another Manu named Danda-dhara Manu in Mahābharata, much earlier than Vaivasvata Manu [Ref. Parva 14, Chapter 4], but later than Chākshusha Manu. However, the names of only nine descendants of Danda-dhara Manu have been given therein. The 9th descendant was king Marutta, who brought gold after getting it extracted from a mine in the North of the Himalayas. This also indicates that advanced technology was available long before Vaivasvata Manu.

References

1. McCrindle, John W., Ancient India as described by Megasthenes and Arrian, 1877
2. Dr. Raja Ram Mohan Roy, Who was 'Sandrakottos', Chandragupta Maurya or Chandragupta I, India Facts.
3. Christopher Lever, Naturalised Birds of the World, 2010, p34
4. Max Muller, A History of Ancient Sanskrit Literature, p743
5. Pushkar Bhatnagar, Dating the Era of Lord Ram, 2004
6. Paul A. LaViolette, Decoding the Message of the Pulsars: Intelligent Communication from the Galaxy, 2006
7. Vedveer Arya, Sanskriti, 2018
8. Michio Yano, in 'The Blackwell Companion to Hinduism', edited by Gavin Flood

9. Egbert Richter-Ushanas, Indu and the Rig Veda, p.16
10. Chandragupta Maurya, New World Encyclopedia
11. Swami Prakashananda Saraswati, The True History and the Religion of India, p.489

GARUDA PRAKASHAN BOOKS

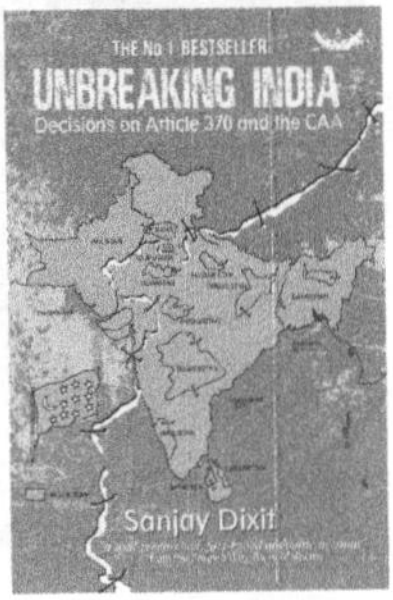

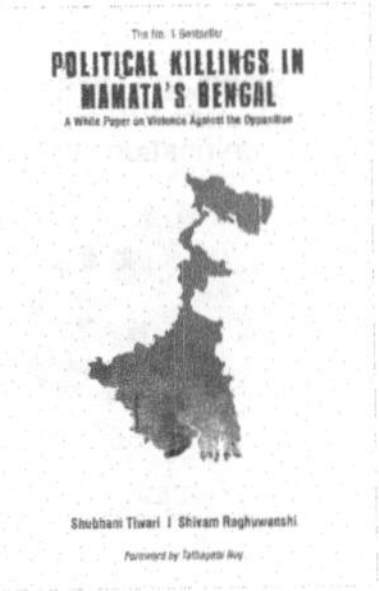

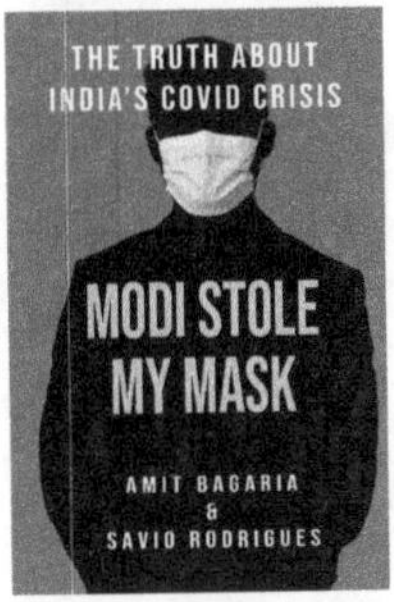

GARUDA PRAKASHAN BOOKS

GARUDA PRAKASHAN BOOKS

गरुड

Register:

Please register your book purchase at **grpr.in/register** to stay in touch and get informed about future books!

To order:

www.garudabooks.com

Follow us:

WEBSITE : www.garudabooks.com

FACEBOOK : www.facebook.com/garudaprakashan/

TWITTER : @garudaprakashan

INSTAGRAM: @garudabooks

YOUTUBE : /garudabooks

Contact:

EMAIL : contact@garudabooks.com

International queries:

EMAIL : international@garudabooks.com

About the Author

Dr M R Goyal has been an engineer by profession, getting his PhD in Engineering and Technology from Punjab University. He retired as Chief Engineer from Punjab Water Resources Department in 1998. Presently, he is Chairman, Board of Consultants for Ranjit Sagar and Shahpur Kandi Dam Projects.

Having keen interest in ancient Sanskrit texts, he has also published a large number of articles in international technical journals.

Dr Goyal has written three books—*History and Reflections of Mahabharata Era,* संक्षिप्त मनुस्मृति and चरक-सुश्रुत-संहिता सार.